FIRE SERVICE
AND THE LAW

FIRE SERVICE AND THE LAW

By
Charles W. Bahme, A.B., J.D.

NATIONAL FIRE PROTECTION ASSOCIATION
470 Atlantic Avenue
Boston, Massachusetts 02210

About the Author

Charles W. Bahme, author of *Fireman's Law Book,* predecessor to *Fire Service and the Law,* has a remarkable record of accomplishment. In thirty years of service with the Los Angeles (Calif.) Fire Department, he advanced through the ranks from recruit to deputy fire chief, the rank he held at the time of his retirement. During that period he obtained his A.B. degree from UCLA and Juris Doctor degree from Southwestern University. Through the same years he served with the U.S. Navy; six years of active duty during World War II and the Korean conflict, and twenty-eight years with the Naval Reserve, in which he held the rank of Captain.

Chief Bahme is an attorney at law with admission to practice in California, before federal district courts, the Supreme Court of the United States, and the highest court of military appeals. He has taught courses in fire protection engineering at UCLA, and fire administration at various state, national, and international conferences. He has served in Europe and the Far East for the Department of Defense and the U.S. State Department.

First Printing
October 1976

Second Printing
April 1977

Copyright © 1976
National Fire Protection Association
All rights reserved

NFPA No. FSP-3A
Standard Book No. 87765-081-0
Library of Congress No. 76-26786
Printed in U.S.A.

Table of Contents

Chapter One **Introduction to the Law** 1
The Meaning of Law, 1
Sources of Law, 1
The Reason for Laws, 4

Chapter Two **Civil Actions** 6
What is a Lawsuit?, 6

Chapter Three **Criminal Actions** 10
What is Criminal Law?, 10
Criminal Prosecutions, 11

Chapter Four **The Judicial System** 16
State Courts, 16
The Federal Judicial System, 18
Court Procedure, 20

Chapter Five **Organization of Fire Departments** 23
Volunteer Fire Departments, 24
Fire Districts, 27
Police-Fire Combination Departments, 34
Are Fire Fighters "Public Officers" or "Employees"?, 35
Are Fire Fighters "Peace Officers"?, 36
Federal Impact on Fire Departments, 37
State Legislative Regulation of Local Fire Departments, 42
Rules and Regulations, 56

Chapter Six **City's Liability for Acts of Fire Department** 63
Conditions Precedent to Municipal Liability, 63
Fire Fighting, a Government Function, 63
Bases for Municipal Nonliability in Tort, 64
Fire Department Liability, 65

v

Fire Apparatus, 70
Mutual Aid and Assistance, 87
Problems in Fire Extinguishment, 89
Fire Hydrants, 94
Torts of Fire Fighters, 96
Other Fire Equipment, 96
Fire Station and Premises, 97
Destruction of Buildings to Prevent the Spread of Fire, 101
Riots—Participation in Civil Disturbances, 102

Chapter Seven   **Municipal Liability to Members of Fire Department**   **106**
Safety in the Fire Service, 106

Chapter Eight   **Fire Prevention Bureaus**   **111**
Source of Power, 111
Fire Prevention Laws, 112
Scope of Police Power, 113
Elimination of Nuisances Constituting "Fire Hazards," 116
Citizens' Personal Liability, 121
Ordinances, 130
Conflict of Jurisdiction, 134
Fire Inspector's Liability, 142
Failure to Enforce Ordinances, 156
Duty of Public toward Fire Inspector, 156

Chapter Nine   **Pensions**   **160**
Pension and Compensation Distinguished, 160
Death or Injury in Line of Duty, 162
Delay in Presentation of Claim, 165
Termination of Employment as Affecting Pension Rights, 166
Ability to Perform Light Duty, 167
Compulsory Retirement Age, 168
Accepting Other Employment, 168
Age Limitations, 169
Disability Pensions, 169
Judicial Determination of Pension Rights, 174

Chapter Ten	**Salary and Compensation** Matter of Law or Contract, 177 Salary, 178 Workmen's Compensation, 182	**177**
Chapter Eleven	**Termination of Employment** Reductions in Force, 188 Removal or Suspension, 188 Necessity for a Fair and Impartial Hearing, 192 Probationary Employees, 196 Demotion, 197 Personnel Included in Removal Safeguards, 198 Reinstatement, 198	**188**
Chapter Twelve	**Duty Owed by Public to Members of Fire Department** Owner-Occupier's Relationship to Fire Fighters, 201 "Firemen's Rule," 204 Application of Statutory Safeguards to Fire Fighters, 205 Application of National Fire Protection Standards to Fire Fighters, 207 Status of Fire Fighters Called Outside of City Limits, 208 Public's Liability for Active Negligence, 208 Duty Owed by Utility Companies, 208 Summary of Public's Duty Owed to Fire Fighter, 210	**201**
Chapter Thirteen	**Liabilities of Fire Fighter** In General, 212 Acting Within "Scope of Authority," 213 Liability as a Public Officer, 214 Discretionary Powers and Ministerial Duties, 218 Liability of Superior Officers, 219 Strikes—What Are They?, 223	**212**

Chapter Fourteen	**Procedural Pointers**	**228**
	Surveys in the Field, 228	
	Authority to Make Arrests, 229	
	Arson, 233	
	Handling "Sit-ins," 236	
	Carrying a Police Badge, 236	
	Signing the Complaint, 237	
	Hearings before the City Attorney, 237	
	Trial Preparation, 237	
	In the Courtroom, 238	
	Epilogue—"But Is It Fair?", 241	
Appendix A	**Visitor's Waiver**	**256**
Appendix B	**Visitor's Identification**	**257**
Appendix C	**California Law on Liability of Public Entities and Employees**	**258**

Foreword

Law is a living and dynamic body of knowledge that responds, however slowly, to the growth of our society. To adequately reflect the changes, it is imperative that new laws be adopted, new judicial decisions be rendered, old statutes be revised, and former judgments be revised or modified. For the most part, the laws, decisions, and administrative orders of regulatory agencies in recent years seem to be casting greater responsibility on the fire service to become more knowledgeable in the application of firesafety principles, and more professional in rendering public assistance. The time is long past when anyone can even jokingly assert that "what you don't know won't hurt you." The standard of care imposed by modern statutes and court decisions that in turn affects the conduct of fire commissioners, administrators, apparatus operators, fire investigators, inspectors, dispatchers, and fire fighters is a matter of serious concern. It demands attention beyond a mere cursory familiarity or vague awareness of the legal implications of actions taken or contemplated in pursuit of the fire service's increasingly complex missions. If nothing else, this book should give the reader a comprehension of the extent the fire service is enmeshed in the fabric of government and law, both historically and in the present social and physical environment.

While many of the new court rulings discussed in this book merely corroborate the more enlightened and progressive decisions discussed in its predecessor, *Fireman's Law Book*, Chief Bahme has attempted to incorporate some of the more important court cases that represent the current judicial trend toward granting fire fighters more freedom of expression and greater participation in fire department administration, while at the same time imposing a higher standard of competence in carrying out duties toward the public. In addition, this book presents practical information about fire protection districts and volunteer fire fighter organizations that Chief Bahme gathered while serving on the board of directors of a fire district following his retirement from active duty as a deputy chief with the Los Angeles Fire Department.

Although the discussions in this study are largely based upon court decisions, the interpretation of the legal effect of certain administrative and judicial rulings in various jurisdictions is sometimes a matter of controversy, particularly where there has been no definitive ruling by the U.S. Supreme Court. Therefore, it should be kept in mind that the opinions given in this book are the private ones of Chief Bahme and are not to be construed as reflecting the official views of the National Fire Protection Association or other organizations of which he is a member.

Chester I. Babcock
Director, Editorial Division
National Fire Protection Association

Preface

This book, the successor to *Fireman's Law Book*, has been prepared as a practical, comprehensive, and up-to-date resource for the members of the fire service. It is designed not merely to give instruction in legal principles applicable to fire fighters and fire departments, but also to give practical information in the application of these principles.

In preparing this book, I have revised the text of *Fireman's Law Book*, making changes and additions to reflect current judicial and administrative thinking on matters of concern to the fire service.

While every effort has been made to check the accuracy of the quotations and references cited in the text, no warranty is expressed or implied, nor is specific counsel intended.

No text can take the place of competent legal advice on a specific problem within a given jurisdiction, but for supplying guideposts on the matters presented it is hoped that this study will be found useful to city attorneys or other legal advisors called upon to render opinions on problems affecting fire fighters or their departments, and especially to members of the fire service.

Charles W. Bahme

Acknowledgments

First I want to express my appreciation for the many helpful suggestions, notes, and citations sent to me by my old friends, Charles Roblee, AB, MA, who has been teaching various fire administration courses at different universities, and to Hollis Williams, a chief officer in the Citrus Heights (Calif.) Fire Department and a member of the California Bar.

I also want to thank Wayne W. Schmidt, senior editor and general counsel for Public Safety Personnel Research Institute, Inc., who kindly suggested that I might freely use any of the many references found in the monthly editions of the institute's *Fire Department Personnel Reporter*.

I also want to acknowledge as a valuable resource some of the interesting cases discussed by Mr. H. Newcomb Morse, LL.B, LL.M, F.A.A.F.C., in his articles prepared for *Fire Command!* magazine.

I also want to thank an old friend and former fellow chief officer, Bob Radke, now retired from the Los Angeles Fire Department and a practicing attorney, for his assistance in researching many of the references cited in the text.

Finally, I must mention my kind neighbor, Melinda Berry, who had the patience to decipher my handwritten scrawls and convert them into a legible manuscript; her help is truly appreciated.

Chapter One
INTRODUCTION TO THE LAW

The Meaning of Law

The word "law" has many meanings, but in its broadest sense it refers to the whole body or system of rules of conduct, including both decisions of courts and legislative acts. Viewed in this light it includes both written and unwritten law, common law and civil law, domestic law and international law, and military law and moral law. The study of law, as presented in the schools, is generally subdivided into about thirty-two major subjects, such as criminal law, torts, real property, equity, contracts, constitutional law, corporations, wills, negotiable instruments, evidence, agency, personal property, domestic relations, pleading and practice, taxation, securities, etc. Because there are so many fields of law, many attorneys become specialists in a particular branch, such as patent law, corporation law, probate law, criminal law, etc., and oftentimes a lawyer who has a general law practice will consult with these specialists when confronted with a difficult case.

Sources of Law

Where is the law to be found? It is not enough to merely refer to the codes or statutes in a particular state to ascertain what the law is respecting a certain problem. The law which governs the conduct of a person is generally to be found in the Constitution of the United States, the Constitution of a state, the statutes of the United States and of an individual state, the ordinances of a city and county, the administrative orders or regulations issued by various governmental regulatory bodies, whether federal, state or local, and in the unwritten law.

Constitutional Law

Unlike England, which has no written constitution, the basic law of the land in this country is the United States Constitution. It is the supreme law, the *organic* law, expressed in written form for determining all private rights and by which all public authority must be administered. Any right which is guaranteed by the Constitution is spoken of as a *constitutional right,* such as the one guaranteed by the Fourteenth Amendment protecting persons against being deprived of life, liberty, or property without due process of law. Such a right cannot be abolished by a state, city, or any other political subdivision, nor by any of its officers or employees. It has been held, however, that some of these constitutional rights, such as the right to bear arms, guaranteed by the Second Amendment, are merely a limitation on Congress, and do not prevent the state legislature from restricting the mode of carrying deadly weapons. Most state constitutions are similar to the Federal Constitution in that all state laws, including local laws and regulations, must not be in opposition to constitutional requirements. In one sense of the word, the charter of a city, granted by a state legislature, is the city's constitution, and acts not only as a measure of the grant of power conferred upon the city, but also as a means of testing the validity of the ordinances adopted by it; a city's ordinances cannot conflict with its charter provisions, nor can such ordinances add or detract from the powers granted by its charter.

Statutory Law

Laws which have been adopted by Congress as federal statutes, and laws which have been passed by the state legislatures are generally termed *statutory law*. When used in its broadest sense, statutory law means *written law* (as distinguished from unwritten law) and includes local ordinances.

Many states incorporate their statutes in various codes. A *code* is a systematic compilation of both statuory law and the law handed down by the judges in their decisions. Typical examples of the state codes in use today are the probate, civil, penal, labor, political, administrative, educational, military and veteran's, health and safety, civil procedure, business and professions, and government code. Local governments also frequently adopt codes relating to firesafety, such as the electrical, building, plumbing, heating and ventilating, refrigeration, and fire codes.

Unwritten Law

The *unwritten law* is often referred to as the *common law,* and also that part of the international law which springs from custom, usage, and ancient practices, as distinguished from treaties. The common law of England consists of a vast number of legal rules distinguished mainly from the Roman or civil law by the fact that these common law rules have never been put in writing and adopted by the legislative body of England. Blackstone's "Commentaries" was, for colonial America, a sort of crystalization of the common law. Thus, the common law, or unwritten law, of England has been used as the basis of our jurisprudence in all the states except Louisiana. The Constitution of the United States contains many of the most valued principles of the common law, as do the constitutions of the various states.

Although many states have adopted the common law by statute, constitution, or decision, its general principles are adopted only so far as they fit particular situations. In states which have adopted codes, the codes are considered a continuation of the common law; however, if the statutes of the state do not agree with the common law, the statutory law is generally held to be paramount.

Many common law offenses have been the subject of laws adopted by statute in the various states, such as the manufacturing or keeping of large quantities of explosives in, or dangerously near, public places, towns, and highways. Such offense is indictable as a common nuisance whether the business is negligently conducted or not.

Summary of Types of Law

In summary, then, law is generally of three types: constitutional law, statutory law, and unwritten law. Sometimes, though, it is difficult to decide exactly what the law is because of its fluid state, influenced by court decisions that continually enter the mainstream. One can predict, however, with a fair degree of accuracy, what the proper rule of conduct is for a given situation by studying the constitution, statutes, and decisions of judges on similar problems. Having done this, and finding no statute or decision directly in point, one can then try to predict what a judge or jury would decide if this particular problem were to be brought before them. The prediction may be wrong and the verdict an adverse one, but if there is still conviction of the rightness of the case and if some possible error in the decision or in the way in which it was determined can be shown, then an appeal can be made to another tribunal. While awaiting this next decision, the law is again uncertain. "The prophecies of what the courts will do in fact, and nothing more pretentious, are what I mean by the law," said Chief Justice Holmes.

Court Decisions

Sometimes the decisions of high tribunals, even of the Supreme Court of the United States itself, may later be reversed by a different set of judges, and the law is changed. Although there are uniform laws on a number of subjects, such as sales, bailments, negotiable instruments, etc., now widely adopted by the various states, the decision of the courts in one state on a given problem may be exactly the opposite of the courts in another state. Sometimes, too, the appellate courts of different districts in the same state do not agree, and until the matter is appealed to a higher court for a final decision, the law on that subject remains uncertain.

It also makes a difference whether a case is brought in a state court or in a federal court; sometimes the subject matter is such that there is a choice of courts, and the attorney will seek to try the case in the court where he is most likely to obtain a favorable decision.

If there is no specific ordinance covering a condition which needs correction, and yet it is necessary to have certain things done under the fire department's general power to secure adequate firesafety, then every effort is made to determine reasonable requirements of the situation; requirements that are likely to be approved by any judge or jury who may later be called upon to review your decision. Until the inspector's statements or orders are actually called into question, for all practical purposes, what he says is the law *is* the law.

As was pointed out in the opening sentence of this chapter, the law is a body of rules governing human conduct. When you direct a man to do a certain thing you are imposing a rule of conduct for him. If you have no authority to direct such action, or such action is not in accordance with statutory requirements, the law which you impose may not be valid, but as far as the man is concerned who acts upon your orders in the belief that what you say is also required by ordinance, your word is law. As a custodian of the law, you have an important trust. Do not violate that trust.

The Reason for Laws

Why do we have laws? A solitary man on a desert island has no need for law. He has no one upon whom to impose any rules that he may care to make, nor is there any commanding authority whose laws he must obey. In a society of people, however, there are many reasons why laws are necessary, even though they result in a restriction of a man's liberty. For example, a man's liberty to swing his fist in the air must stop at the point where it is likely to hit another man's nose; otherwise, there is likely to be a fight or other breach of the peace. Among the objectives which the law seeks to accomplish are the following:

1. To eliminate friction among people and thus prevent disturbances of the peace.

2. To protect the personal rights of individuals, such as freedom from injury, rights of assembly, freedom of speech, etc.

3. To protect the property rights of people, both as to personal belongings, and as to real estate.

4. To establish standards of conduct so that a man may know what course of action is expected of him under given circumstances, and at the same time be able to predict what the other fellow is likely to do in the same situation. For example, the safe operation of motor vehicles is dependent upon the above objective.

5. To provide remedies for the invasions of a man's rights, such as injuries to his person or property.

6. To provide for the punishment of persons who do not respect the rights of others, and thereby discourage antisocial behavior which might affect the well-being of the public as a whole.

To carry out the fifth and sixth objectives listed above, a vast body of law has been developed throughout the ages. As a part of this introductory study of the law, we shall consider next the two principal kinds of actions dealt with by the courts: civil actions, or ordinary lawsuits, and criminal actions, or prosecutions.

Review Activities

1. With a group of your classmates, discuss the differences between statutory law, common (unwritten) law, and constitutional law. Then, in your own words, write a definition for each. Compare your definitions with those of

your classmates and, if necessary, revise your definitions to include any constructive suggestions.
2. Most state constitutions are similar to the Federal Constitution in that all state laws, including local laws and regulations, must not be in opposition to constitutional requirements. Therefore, the charter of a city, granted by a state legislature, can be considered to be a city's constitution; as such, it can be said to act as a measure of the grant of power conferred upon the city. How might the charter of a city also affect the city's ordinances? Why?
3. For what reasons might an unfavorable verdict be appealed to another tribunal?
4. Explain why, for most of the people advised on fire prevention matters, the fire inspector's word is usually accepted as the final authority.
5. With two or more of your fellow students, discuss the reasons for laws and some of the objectives the law seeks to accomplish. Then list at least five one- or two-sentence examples, other than the one given in this chapter, that illustrate why laws are necessary even though they result in a restriction of personal liberty. Next to each of your examples, write the number of the objective (1 through 6 from the last section in this chapter) sought by the law that best matches the example. Compare your examples with those of your classmates, and discuss the appropriateness of the objectives as matched with the examples.
6. Based on what you have learned in this chapter, write a one-paragraph statement summarizing what the word "law" means to you.

Chapter Two
CIVIL ACTIONS

What Is a Lawsuit?

A *lawsuit* is a proceeding in a court, started by one party to compel another to do him justice. *Justice* is probably indefinable, for what may seem just to one man may seem altogether unjust to another man. A lawsuit is a civil action and not a criminal prosecution. You can usually tell by looking at the name of a case whether it is a report of a civil action or of a criminal action.*

The parties in a lawsuit are the plaintiff and the defendant. The *plaintiff* is the one who causes the action to be brought against the other person; in the footnote below John Doe is the plaintiff and Richard Roe is the *defendant,* for he is the party being sued. You may sometimes see the parties to a lawsuit referred to by other names. If the case is appealed to a higher court, the party who appeals is called the *appellant.* It makes no difference whether the party who appeals was the plaintiff or the defendant in the lower court. He becomes the appellant in the higher court and the party against whom the appeal is taken is called the *respondent.*

The object of a lawsuit is usually to seek damages for some injury caused the plaintiff by the defendant. For example, in a libel or slander suit redress is sought for injuries caused by the defamation of the plaintiff's character; such injuries may have caused him to be subjected to ridicule, loss of prestige, loss of business, etc. If such defamation was made in writing the action is called *libel,* and if made orally, is called *slander.* Examples of other civil suits include actions to recover damages for breach of contract, negligent operation of vehicles, etc. Some civil suits do not have as their object the obtaining of money damages, but seek some other remedy, such as the reformation of a contract, the return of property, the probating of a will, etc.

Distinction between a Civil and a Criminal Action

Sometimes a civil suit and a criminal prosecution can be maintained for the same act, though not in the same trial. For example, if a man so unlawfully

---

* For example, if you see a reference which reads like this: John Doe v. Richard Roe, 214 Cal. 241, 4P. (2d) 929 (1931), it means that John Doe sued Richard Roe; *v* means *versus* (against); the lawsuit (case) is set forth in volume 214 of the *California Reports* at page 241; the last part of the citation indicates that the case is also reported in another set of books called the *Pacific Reporter,* in volume four of the second edition, on page 929. To *cite* a case means to refer to it, as in the example above, or to quote it as authority for a principle of law. The example given above is a civil case, for one man is suing another, and the state is not a party.

threatens you in such a manner as to cause you to fear that he is about to do you bodily harm, he can be prosecuted for assault. If he actually carries out his threat, the crime of battery is also committed. If you were to have him arrested and prosecuted, the *state* or the *people* would be the prosecuting party in the criminal action, the person committing the act would be the defendant, and you would be the *complainant,* or prosecuting witness. But in addition to the criminal action, you could file a civil action (lawsuit) against the man to recover damages for any injuries resulting from either the assault or battery or both. In the civil action you would be the plaintiff and he the defendant. You could not use the verdict against the defendant in the criminal case to help prove your case in the damage suit or vice versa; you could not even bring up the fact that there had been a criminal prosecution unless the defendant makes statements contradictory to the previous trial, and then this contradiction could only be brought out to impeach the credibility of the defendant.

In a criminal action the primary objectives are to punish the defendant for his wrongful acts and to furnish an example which will discourage others from acting in a similar manner. In criminal actions, the state is the party principally interested in seeing that justice is carried out. In a civil action the primary objective is to seek some remedy, such as damages, a divorce, performance of a contract, an injunction, etc., and the private individual—the plaintiff—is the party principally interested in the outcome of the action rather than the state.

A word of warning: Never threaten criminal prosecution against a man who has caused injury as a means of getting him to pay damages. Accepting money from a man who has committed a crime on the promise that there will be no complainant for the state in a prosecution is in itself an offense known as *compounding a crime*. For a public official to conceal a crime, or even to refuse to enforce a law that he is duty bound to enforce, in exchange for some benefit to himself is also a crime, known as *bribery*. Take a case where there is no question of official duty involved, such as might occur when a man has stolen a car from you. It is all right for you to promise not to sue him for damages in a civil action if he will return it to you, and you are not under obligation to go down to the nearest magistrate and file a criminal complaint against him for theft. In fact you may accept payment for injuries and restitution of property without ever seeking legal redress of any kind. But, if someone else should file a criminal complaint against him for having committed the theft, and you are later called upon to testify in court as to what took place, you must tell all you remember about the incident and not let your memory become vague merely because the man settled matters to your satisfaction. It is the state which must be satisfied in criminal matters and not the individual; however, the court may modify the punishment in cases where the defendant has made restitution and redress before being brought to trial. (Criminal actions are further discussed in the next chapter.)

Who Can Sue in a Civil Action?

Anybody can bring a lawsuit against anyone who injures him, whether the injury is intentionally committed or negligently inflicted. Of course it is necessary for children (or unborn infants) and people who are mentally incompetent to have someone bring the suit for them; their guardians do this. This does not mean, however, that every time a person is injured he can sue someone. There are times when no one can be held legally responsible for the injury; such situations are referred to as *damnum absque inuria* (an injury without a wrong). For example, an inspector, acting within the scope of his authority, directs a plant operator to install a wire screen in front of his infrared drying lamps in the belief that it will protect the lamps from being broken by accidental contact with freshly painted metal parts passing in front of the lamps. The fact that the screen, installed at considerable expense, absorbs most of the heat from the lamps and renders the drying process useless, is a wrong for which the law provides no legal remedy. In making the recommendation the inspector was acting within the scope of his authority. If, with no malice, but in good faith on his part, he recommends something that turns out to be impractical, the recipient of his advice is without legal recourse.

Who May Be Sued in a Civil Action?

Anybody may be sued in a civil action. Even children may be sued for their torts. A *tort* is an injury or wrong committed, either with or without force, to the person or property of another; such injury may arise by the nonfeasance, malfeasance, or misfeasance of the wrongdoer. *Nonfeasance* is the failure to perform a duty, such as one owed to the public; *malfeasance* is the doing of an act which a person ought not to do at all; and *misfeasance* is the improper doing of an act which a person might lawfully do.

There are many limitations on the right to sue, such as a time limit in which to bring the action (determined by statutes of limitations), or the place in which the action must be brought (determined by *venue* statutes), and in the case of the United States, compliance with the restrictions under which permission to sue it is granted by Congress.

Telling a man that he cannot sue you is not a statement of fact. He may not have a good cause of action, but that doesn't keep him from suing you. If he is without a good cause of action he will probably fail to obtain a judgment against you, but if he sues you anyway you may find it necessary to retain a lawyer for your defense to prevent a default judgment from being rendered against you.

Furthermore, if a man does sue you without *probable cause* (reasonable grounds) you can later sue him for malicious prosecution, and you may obtain a judgment against him for damages. But even if you obtain a favorable decision against him, it does not necessarily mean that you will collect your damages. He may have no money or property, or he may be exempt from execution or attachment, and therefore you may wind up with practically nothing but a moral victory—even if you win. A word of advice: *it pays to investigate before you litigate.*

The prime requisite for a successful lawsuit is a good cause of action. A *cause of action* arises when a legal right of the plaintiff is invaded by some action of the defendant. It is a *good* cause of action when the party being sued lacks a good defense. Even though the plaintiff's rights have actually been invaded by the defendant, he may not win his lawsuit for several reasons. The defendant, in his pleading (called an *answer*) to the plaintiff's declaration (called a *complaint*), may present several defenses. He may deny that the plaintiff ever had a cause of action, or, if he had, that he has none now. (These defenses are known as *pleas in bar* at common law.) He might also raise such objections as lack of *venue* (action brought in the wrong county), or lack of jurisdiction of the court, e.g., where the defendant was never properly served with a summons, or incapacity of the plaintiff to sue, or pendency of another action on the same matter. These defences are known as *pleas in abatement* for they set forth facts which merely show that the plaintiff does not at the present time have a good case because of some technicality, and therefore the action should be abated, or held up, until the defects are corrected.

Another type of defense is one which admits the truth of the plaintiff's allegations, but introduces new matter to avoid the effect of his claims and to show that the plaintiff is not entitled to recover any damages. For example, the defendant may admit that he regularly drove his auto over the plaintiff's land, and may then produce a written license from the plaintiff which proves that he had the right to cross over it. Or, for example, he may admit that his auto collided with the back of the plaintiff's car, but then claim that he could not avoid it because the plaintiff backed his vehicle without looking behind and observing the defendant's car lawfully parked at the curb. Such pleas in common law are called *pleas of confession and avoidance.*

Review Activities

1. You have been sued without probable cause. In turn, you sue for malicious prosecution and obtain a favorable decision for damages. For what reasons might you not be able to collect damages? Should one of these reasons be applicable to your case, you may have achieved a moral victory. Explain, by presenting an example, how a moral victory can be important.
2. What is the distinction between a civil action and a criminal action? Give an example of each.
3. Write one-sentence descriptions for each of the following terms:
    (a) slander   (d) lawsuit   (g) respondent
    (b) plaintiff   (e) apellant   (h) libel
    (c) cite a case   (f) complainant   (i) defendant
4. Explain the meaning of "compounding a crime" as used in criminal prosecution. Give an example.
5. As a fire inspector acting within the scope of your authority, you direct the manager of a paper products warehouse to move stored products from an unsprinklered area to a sprinklered area. A week after the move, a sprinkler accidentally operates and damages the materials. Does the manager of the warehouse have legal recourse? Explain why or why not.
6. Necessary to the success of a lawsuit is a *good cause of action*. In your words define "cause of action" and "good cause of action," including in your definition appropriate examples of each.

Chapter Three
CRIMINAL ACTIONS

What Is Criminal Law?

Criminal Law is the branch of jurisprudence that deals with crimes and offenses. Anyone choosing to live as a member of society rather than as a hermit must give up part of his liberty, and submit to criminal laws which, in certain cases, authorize the imposition of penalties, the deprivation of liberty, and even the destruction of life, with the view to the future prevention of crime and to protect the safety and welfare of the public. It is important to have a working knowledge of criminal law, for ignorance of the law is no excuse for its violation. Everyone is presumed to know the law, and this is true even if the statute making an act illegal is so recent as to make it impossible to practically know of its existence. Although the definitions of crimes, the nature of punishments, and the forms of criminal procedure have their origin largely in the common law, many of the rules concerning crimes have been modified by statutes such as are commonly found in the penal codes of the various states.

Some of the important principles of the American and English system of criminal law are:

First   Every man is presumed to be innocent until proved guilty, and he is entitled to any benefit of the doubt where any reasonable doubt exists concerning his guilt or innocence.

Second   No person can be brought to trial until there has been a proper indictment by a grand jury (though in some states, as in California, the alternative procedure of filing of an information or a complaint is now permissible).

Third   The accused is entitled to a trial by a jury of his peers (equals in rank), selected impartially from among the people, and whose decision on questions of fact is final.

Fourth   In contrast to the European system, he must be tried on the basis of the facts presented and not on his character and his prior conduct.

Fifth   The prisoner cannot be required to give testimony which will *incriminate* him (subject him to possible punishment for the commission of a crime). This rule does not mean, however, that an accused who has testified in his own behalf cannot be made to answer questions on cross examination.

Sixth He cannot be tried for the commission of an act which was not a crime at the time he committed it, nor can the punishment imposed be increased over what it was at that time. (A law which violates these rights is an *ex post facto law*.)

Seventh The accused cannot be put in jeopardy twice for the same offense. (The fifth amendment to the Constitution of the United States forbidding double jeopardy only prohibits double punishment for the same offense in the federal courts, but does not forbid the states from also punishing for the act when the same act may be considered a separate and distinct offense against the state; nor does it prohibit either the state or federal government from passing a law authorizing the separate punishment of each step leading to the consummation of the crime and also punishment of the completed crime.)[1]

Criminal Prosecutions

A criminal action is a prosecution in a competent court of justice in the name of the government for the commission of a crime. A *crime* is a wrong considered of public character because it possesses elements of evil which affect the public as a whole and not merely the person whose rights or property have been invaded. By statute in many states a *crime* or public offense is specifically defined as an act committed or omitted in violation of a law forbidding or commanding it, and to which punishment is imposed upon conviction. Punishments in California, for example, range from death, imprisonment, fine, removal from office to disqualification to hold or enjoy any office of honor, trust or profit in the state.[2]

Crimes are usually divided into two categories: felonies and misdemeanors. At common law, a felony was any crime which, like treason, worked a forfeiture of the offender's lands or goods. In modern practice the distinction between a crime and a misdemeanor is made to depend upon the character of punishment provided by statute. As a general rule all crimes punishable by death or imprisonment in a state prison are *felonies*. Every other crime is a *misdemeanor*. In about a dozen states, there are special provisions relating to juveniles, such as those applied in the juvenile courts of California under the provisions of the California Youth Authority, whereby a minor who has been committed to that authority may apply to the court to have a felony of which he has been convicted changed to a misdemeanor upon his discharge.

In every crime or public offense there must exist a union, or joint operation, of act and intent, or criminal negligence. Such intent is shown by the circumstances connected with the offense, as well as the soundness of mind and the discretion of the accused. All persons are of sound mind who are sane. The legal and medical ideas of insanity are essentially different. Insanity in medicine

---
[1] Albrecht v. United States, 273 U.S. 1, 47 S. Ct. 250 (1927).
[2] West. Cal. Penal Code, ¶15 (West 1970).

has to do with the long departure of the individual from his natural mental state arising from disease, whereas the legal problem must resolve itself into the inquiry whether there was mental capacity and moral freedom to do or abstain from doing the particular act. A man can be insane in the medical sense and still be held accountable for his criminal acts. For example, a man may have the insane delusion that he can talk with horses or fly like the birds, or that he is Moses reincarnate, and yet understand perfectly the difference between right and wrong. If such a man were to deliberately murder another man under circumstances where his particular delusion had no bearing on the killing, then he could be found sane insofar as his trial for murder is concerned. Likewise, if he were a medical paranoiac (one with a systematized delusion of persecution), and if he had the delusion that a certain man had done him an injustice, and if he were to poison his food, he might nevertheless be found legally sane and prosecuted on the theory that he knew that a desire for revenge is no justification for homicide.

Criminal negligence is the neglect of a legal duty imposed either by statute or by the common law; however, in those states that have abolished the common law the bare neglect of a legal duty is not a crime unless some statute so prescribes. A *criminal offense,* as used with reference to the introduction of evidence in a trial to impeach the credibility of witnesses (by showing previous conviction of criminal offenses), does not include the violation of municipal ordinances, though the term does embrace both felonies and misdemeanors.

Who Is Capable of Committing Crimes?

It is generally conceded (specifically in some states) that all persons are capable of committing crimes except those belonging to the following classes:

1. Children under the age of fourteen who, in the absence of clear proof that at the time of commiting the act charged against them, knew its wrongness.

2. Idiots, lunatics, and insane persons.

3. Persons who committed the act or made the omission under an ignorance or mistake of fact sufficient to disprove any criminal intent.

4. Persons who committed the act charged without being conscious of doing so.

5. Persons who committed the act or made the omission through misfortune or by accident, when it appears that there was no evil design, intention, or culpable negligence.

6. Married women (except for felonies) acting under the threats, command, or coercion of their husbands.

7. Persons (unless the crime be punishable with death) who committed the act or made the omission because of threats or menaces sufficient to show that they had reasonable cause to, and did, believe their lives would be endangered if they refused.

Who Is Liable to Punishment?

It is generally (specifically in some states) held that the following persons are liable to punishment under the laws of a given state:

1. All persons who commit, in whole or in part, any crime within the state under whose laws they are being punished.

2. All who commit robbery, larceny, or embezzlement (as defined by the state in which arrested), though the crime was committed entirely outside the state, but who bring the stolen property, or any part of it (or if found with it) into the state.

3. All who, being outside the state, cause or aid, advise or encourage another person to commit a crime within the state, and are afterwards found therein.

Drunkenness is no excuse for crime. No act committed by a person while in a state of voluntary intoxication is less criminal by reason of his having been in such a condition. But whenever the actual existence of any particular purpose, motive, or intent is a necessary element to constitute any particular crime (or degree of a crime), the jury may take into consideration the fact that the accused was intoxicated at the time in determining the purpose, motive, or intent with which he committed the act.

Parties to Crime

Parties to crimes are in two main categories, principals and accessories. A *principal*, within the meaning of criminal law, is any person who is present at the scene of a crime and who participates directly or indirectly in its commission. By statute in some states the term has been enlarged to include a person who though not actually present at its commission, has advised and encouraged it, and including any person who gets another drunk for the purpose of causing him to commit a crime, or who uses threats or coercion to get him to do it.

An *accessory* is a person who in some manner is connected with a crime, either before or after it is committed, but who is not present at the time it is perpetrated. In early law, accessories were divided into three classes: *accessory before the fact* (who though not present, encouraged the crime to be done); *accessory at the fact* (who was present at the time, but only as a look-out or assistant); and *accessory after the fact* (who, knowing that a crime has been committed, assists the perpetrator to escape, or who hides, comforts or assists him in any way to avoid arrest). In states like California, accessories before the fact and at the fact have all been put in the same class as principals, and all others (who later harbor the criminal) are called accessories.

Types of Crimes

Although crimes are sometimes classified according to the degree of punishment incurred by their commission, they are more generally arranged according to the nature of the offense. The following, based upon Bouvier's *Law Dictionary and Concise Encyclopedia* analysis, is perhaps as complete a classification as is warranted here:[3]

1. **Crimes against the sovereignty of the state**
   e.g., treason (giving aid to one's country's enemies), and misprision of treason (knowledge and concealment of treason without otherwise consenting to the crime).

---
[3] 1 Bouv. 729, 730.

2. **Crimes against the lives and persons of individuals**
   e.g., murder, manslaughter, mayhem, rape, robbery, kidnapping, false imprisonment, abduction, assault and battery, abortion, cruelty to children.
3. **Crimes against public property**
   e.g., burning, destroying, or injuring public property.
4. **Crimes against private property**
   e.g., arson, burglary, larceny, embezzlement, malicious mischief, obtaining goods by fraud.
5. **Crimes against public justice**
   e.g., perjury, bribery, destroying public records, counterfeiting public seals, jail-break, escape, resistance to officers, obstructing legal process, barratry, extortion, contempt of court, suppression of evidence, compounding a crime.
6. **Crimes against the public peace**
   e.g., unlawful assembly, riot, challenging or accepting a challenge to duel, breach of the peace, libel.
7. **Crimes against chastity**
   e.g., adultery, bestiality, buggery, bigamy, incest, seduction, fornication, pimping, prostitution, keeping or frequenting a house of ill-fame.
8. **Crimes against public policy**
   e.g., lotteries, gambling, immoral shows, obstructing the right of suffrage, destruction of fish and game, nuisance.
9. **Crimes against the currency and public and private securities**
   e.g., counterfeiting money, passing counterfeit money, forgery.
10. **Crimes against religion, decency and morality**
    e.g., blasphemy, profanity, Sabbath-breaking, obscenity, cruelty to animals, drunkenness.

Although some of the examples given above may no longer be considered crimes in some states, such as lotteries, Sabbath-breaking, etc., in those jurisdictions where they are illegal, they would fall in the classification indicated; the examples given do not purport to include every act which could be deemed criminal in one or more states.

Review Activities

1. Write your own definitions of the terms, "felony" and "misdemeanor." Include in your definitions an example of each, as related to the public fire department.
2. An arson suspect has been apprehended. However, the suspect is mentally incapable of remembering having set the fire. Would this suspect be considered capable of having committed the crime? Why or why not?
3. In a written statement, explain the difference between an *accessory* and a *principal* as parties to a crime.
4. With a small group of classmates, discuss and classify by types the following three crimes: (1) arson, (2) malicious mischief, and (3) burning. Then,

based on your own experience or knowledge, write a one-paragraph example of each.
5. Explain the differences between *criminal negligence* and *criminal offense*.
6. Read the following paragraph. Based on what you have learned in this chapter, explain why the statements lettered (a) through (e) are true or false. Defend your explanations by citing specific passages from Chapter 3.

Because your community has a high percentage of false alarms, devices for filming the person(s) activating an alarm have been installed on fire alarm boxes. During a false alarm, one such device took a picture and the police have apprehended a suspect.

(a) Because the suspect is on film, he is automatically presumed to be guilty.
(b) The suspect, who has been brought to trial and has testified in his own behalf, cannot be cross examined.
(c) The suspect can be brought to trial without a grand jury indictment.
(d) The suspect is tried on the basis of his good character and prior excellent conduct.
(e) The suspect, having been exonerated, can again be brought to trial on the basis of new evidence not presented at the first trial.

Chapter Four
THE JUDICIAL SYSTEM

In actuality there are two judicial systems: the state court system and the federal system. Whether a case is brought in the state courts or in the federal courts is a matter of jurisdiction. Jurisdiction is of two kinds: over the person, known as *in personam;* and over the thing, or *in rem.* Jurisdiction of both kinds is exercised by both the state and federal courts within the territorial limits of their authority to try cases. The court's *jurisdiction,* i.e., its authority to hear cases, is also limited by the nature of the case sought to be tried. Jurisdiction is also classified as either original or appellate. The court in which a case must first be tried is one of *original jurisdiction,* at least as far as that particular subject is concerned. If the court has the power to hear and determine appeals from lower courts it is a court of *appellate jurisdiction.*

For example, a man may wish to sue a debtor in Los Angeles to collect $2,500 owed him. He must commence this action in the municipal court, for it has jurisdiction over cases involving amounts up to $5,000; an appeal from that court's decision would be made to the superior court (called supreme court in some states, such as New York), for it is the highest court of appeal on such matters from a municipal court.

Some courts have both original and appellate jurisdiction; the very lowest courts in the judicial system have original jurisdiction only, while the rest have original on some matters and appellate on others.

State Courts

The judicial system of a state is comprised of a supreme court, district courts of appeal, superior courts (sometimes called "supreme courts"), municipal courts, and such inferior courts as the legislature may establish; in addition, the state senate is usually empowered to try impeachments of members of the legislature.

The Judicial Council

A majority of the states have adopted some form of judicial council, which is a body set up to study a state's judicial system, to compile statistics, to do judicial research and to recommend improvements.

The Supreme Court of the State

The supreme court is the highest court of appeals in the state. In California it is composed of a chief justice and six associate justices. It may sit as a group or may split up into two departments of three justices each. The supreme court has presented to its cases appealed from the superior courts, usually on matters involving equity jurisdiction and cases in law relating to real estate, taxes, and probate matters. If the death penalty is involved, it may hear appeals of criminal prosecutions. The supreme court also hears appeals from the district courts of appeal. It has the power to issue *writs* necessary to exercise its jurisdiction; such writs include *habeas corpus, mandamus, certiorari,* etc. (See Appendix A for definition of these terms.)

District Courts of Appeal

A state may be divided into several districts, comprised of a number of counties in each one, for the purpose of setting up district courts of appeal. These courts have jurisdiction over appeals from the superior courts which do not go directly to the supreme court of the state. In California there are five districts, and some of these have more that one division. For example, in District II, where the court meets in Los Angeles, there are five divisions, with three justices in each division.

Superior Courts

The important trial courts in a state's judicial system are its superior courts, or, by whatever name designated (e.g., in New York, "supreme court"), those courts which have mainly original jurisdiction in civil matters, including some tax cases, probate (relating to the estates of deceased persons), and domestic relations cases. Juvenile cases usually go directly to the superior court, and so do petitions for such remedies as habeas corpus and injunctions. The superior court also has jurisdiction over prosecution for criminal actions involving felonies and certain misdemeanors. Larger counties have separate appellate departments of their superior courts.

Municipal Courts

Cities of large population will generally have municipal courts to handle civil cases not involving large sums of money and all misdemeanors except those committed by juveniles.

Inferior Courts

Cities having a municipal court (which includes a small claims court) do not usually have other inferior courts. Such courts include (1) township justice courts, (2) city justice courts, (3) inferior "municipal" courts such as "recorder's" courts and "police" courts in some towns, (4) police courts, and (5) city courts.

The delineation of jurisdiction in inferior courts has not always been clear. For example, a constitutional amendment was passed in 1950 to clear up the confusion in the lower courts of California. Prior to that time the judge in a

Class B Justice Court was not even required to be an attorney, and in some police courts the judge was actually hired and fired by the chief of police. Under the new law the only courts inferior to the superior courts are the municipal and justice courts.

Reports of State Courts

The decisions of the lower courts are not usually reported in cumulative publications as are those of the courts of appellate jurisdiction; the latter are published under various titles. In California, for example, the reports of the cases decided by the Supreme Court are called the *California Reports*, while the reports of the cases decided in the District Courts of Appeal are called *California Appellate Reports*.

In addition to the official publications of the various states, such cases are also published by a private company in what is called the *Reporter System;* for example, the *Pacific Reporter* covers the cases decided in a number of western states, while the *Atlantic Reporter* covers a number of the eastern states.

The Federal Judicial System

The President appoints all the judges of the federal courts. With the exception of the judges of the Tax Court, who have a twelve-year term, they hold office for life and may retire with full pay at the age of seventy.

The United States Supreme Court

The highest court of the United States is the Supreme Court. Though it has original jurisdiction in some cases, such as those where one state sues another, most of its work involves the settling of appeals taken from the highest state courts (where a federal question is in issue), and in hearing appeals from the Circuit Court of Appeals; occasionally an appeal is heard direct from the federal District Courts.

The Constitution of the United States does not expressly grant the Supreme Court of the United States the authority to declare a law of Congress or that of a state invalid, but in 1805 it asserted that it had this power in the famous case of Marbury v. Madison.[1] Chief Justice Marshall based his reasoning on the premise that since the constitution is the supreme law of the land, it is up to the highest court to hold invalid any attempt to circumvent its provisions; to hold otherwise, it declared, would nullify the value of a written constitution and permit it to be altered by any statute which the current congress or state legislature happened to pass.

Federal District Courts

The federal district court has broad jurisdiction, including all matters which come under the postal, bankruptcy, and federal banking laws. It has jurisdic-

---
[1] 1 Cranch 137 (1803).

tion over all cases where the amount in question is over $10,000 and requires interpretation of a federal law or treaty, or the U.S. Constitution, or where the dispute involves citizens of different states. It also has jurisdiction over all cases of admiralty and all cases of crimes punishable under the laws of the United States or committed on the high seas.

Each state has at least one court; but many states have two or three districts; California, Texas, and New York have four districts each. A district itself may be divided into divisions and may have several places where the court hears cases. Only one judge is usually required to hear and to decide a case in a district court, but in some kinds of cases it is required that three judges be called together to comprise the court.

The Circuit Court of Appeals

The circuit court of appeals hears appeals from the federal district courts. The United States is divided into 11 federal judicial circuits in each of which there is a court of appeals. Each circuit includes three or more states, except the District of Columbia Circuit. The states of Alaska and Hawaii and the Territory of Guam are included in the Ninth Circuit, Puerto Rico is included in the First Circuit, the Virgin Islands in the Third Circuit, and the Canal Zone in the Fifth Circuit. Each court of appeals usually hears cases in divisions of three judges.

The United States Court of Claims

The United States Court of Claims has jurisdiction over all claims made against the federal government except in the matter of pensions. This court was established to overcome the traditional ban against a citizen's suing the sovereign authority. Because of the ancient assumption that the "King could do no wrong," it has long been forbidden to sue the government without its consent. Congress has enacted laws which set up the conditions under which the federal government consents to be sued, and it also makes the necessary appropriations with which to pay any judgments rendered against the United States. The court of claims has nationwide territorial jurisdiction. Trials for the purpose of taking testimony and receiving exhibits are conducted before commissioners at locations most convenient for the claimant and his witnesses.

The United States Court of Customs and Patent Appeals

The United States Court of Customs and Patent Appeals hears appeals from the decisions of the patent office on matters concerning trade-marks and patents, and also hears appeals from the United States Customs Court; the latter has jurisdiction to decide disputes involving tariff rates, revenues, appraisals, classification of imported goods, etc.

The United States Tax Court

The United States Tax Court hears appeals from the deficiency tax assessments made by the Commissioner of Internal Revenue on taxpayers who have not paid enough income or estate taxes. It does not handle claims for overpayment of taxes; the latter are presented in the U.S. Court of Claims or the U.S. District Court; they must be brought in the latter court if the claim exceeds $10,000.

The United States Court of Military Appeals

Although it operates as a part of the U.S. Department of Defense for administrative purposes, the court is judicially independent. It works jointly with the Judge Advocate General of the armed services and the General Counsel of the Department of Transportation, and reports annually to the Congress on the progress of the military justice system. It has jurisdiction over cases involving high ranking military officers, those extending to the death sentence, and those accepted or certified after a review by a court of military review.

Court Procedure
Civil Suits

The principal steps in a civil suit are:
1. A complaint is filed with the court by the plaintiff.
2. A summons is issued by the court which gives the defendant a certain number of days to answer the complaint.
3. The defendant must file a pleading within the prescribed time; this may be an "answer" or a *demurrer*. Where a demurrer is interposed the court rules on the question of law as to the sufficiency of the pleading.
4. A trial is conducted before a judge or a jury of twelve people, though the constitution of some states permits less than twelve if the parties agree.
    (a) Plaintiff's attorney opens the case by calling witnesses and presenting evidence to prove his contentions.
    (b) Defendant's attorney cross-examines plaintiff's witnesses in an attempt to bring out facts damaging to plaintiff's case.
    (c) Defendant's attorney then puts his witnesses on the stand, and they in turn are cross-examined by the plaintiff's counsel.
    (d) Attorneys give their closing arguments if they so desire. (Attorney for the plaintiff has the right to give the first and last of such arguments.)
5. The judge instructs the jury in matters of law; the jury decides all questions of fact in the case, and applies the law as instructed. If the trial has been conducted without a jury, the judge takes the case under advisement.
6. The jury brings in a *verdict;* at least three-fourths of the jurors must be in agreement in a civil matter. When there is no jury the judge renders a judgment.
7. There may be a request for a new trial because of some serious error committed during the process, or an appeal may be made to a higher court because of some errors in law, or error in judging the evidence.
8. If the plaintiff is successful he will attempt to collect the amount set forth in the judgment or verdict. This may require the levying of an *execution* against the defendant's property, whereby the property may be seized by the sheriff and sold to satisfy the amount awarded.

Criminal Procedure

The principal steps in the prosecution of a criminal action are as follows:

1. If a citizen comes to the city attorney (or county counsel or district attorney, if in the county) to protest the acts of a neighbor, the attorney may have a complaint drawn, and if for a misdemeanor, send out a notice to the violator to appear in court for arraignment at a certain time; if the matter is serious and requires immediate action, he can have a bench warrant issued for the offender's arrest and have a peace officer take the person in custody. For certain crimes an arrest may be made by an officer or citizen with or without a warrant (see Chap. 14).

2. The accused is brought before a judge for a preliminary hearing. If the judge believes the evidence is sufficient, he can have him put in jail or require him to furnish bail for his release. This *bail* may consist of a cash deposit or a written guarantee of his appearance in court at the proper time by a licensed bonding company or by persons who own sufficient property locally to be financially trustworthy.

3. A man is usually charged with a crime by information or indictment. An *information* is a complaint, filed by the district attorney, public prosecutor, or city attorney, based upon facts brought out at the preliminary hearing, and must be examined by a judge before being brought into court. An *indictment* is the accusation of a grand jury before whom the facts have been presented. Each county in a state is usually required to empanel at least one grand jury per year.

4. The person charged with the crime pleads "guilty" or "not guilty" at the stage in the process known as *arraignment for pleading*. If he pleads "guilty," he is sentenced then and there. If he pleads "not guilty," a date for trial is set.

5. At the date set for trial, unless waived by the accused, a jury is selected to try his case.

6. The trial proceeds as follows:
    (a) Prosecutor opens his presentation and produces witnesses and evidence to prove his case.
    (b) Defendant's attorney cross-examines the complaining witnesses. The prosecutor may then reexamine them if he desires.
    (c) Attorney for the accused then puts on the case in defense.
    (d) Prosecutor cross-examines defendant's witnesses, after which the defendant's counsel may reexamine them.
    (e) Defendant's attorney and the prosecutor give their closing arguments.

7. Judge charges the jury and instructs them as to the law which should be applied in the case; he may also comment on the evidence and the witnesses.

8. After considering all the evidence and applying the law contained in the written instructions handed them, the jury announces its verdict. All of the jurors must agree or else there can be no conviction; if even only one juror dissents, there is a "hung jury," and the accused must be released or be tried again. If a verdict of "guilty" is found, the judge gives the sentence.

9. The accused may appeal the verdict or ask for a new trial on various grounds which tend to show that he did not have a fair trial. The state, however, cannot appeal if the accused is found "not guilty."

In 1972 the Supreme Court of the United States upheld the validity of a Florida law which permitted six-man juries in all criminal cases not involving the death penalty.[2] Twelve other states also provide for juries of fewer than twelve members in criminal cases.

Review Activities

1. In outline form present, in your own words, the principal steps in a civil suit.
2. What types of cases would be termed "criminal actions"? Explain what happens if the criminal pleads guilty. Then explain what happens if the criminal pleads not guilty.
3. Describe the procedure for the prosecution of a criminal action by creating an imaginary situation and, on a step-by-step basis, applying to it the principal steps involved.
4. To which particular court would each of the following cases be assigned?
    (a) A discrepancy in an income tax.
    (b) The settling of a will.
    (c) The establishment of a trademark.
    (d) The State of Ohio suing the State of Arizona.
    (e) Filing for bankruptcy.
    (f) Disputing a $10 parking ticket.
    (g) Suing a person for $3,500.
    (h) A case involving a fourteen-year-old boy.
5. In the United States there are two judicial systems: (1) the state court system, and (2) the federal system. Write your own definition of each.
6. Explain, by citing examples, the differences between the Circuit Court of Appeals and the Court of Customs and Patent Appeals.

---

[2] Colgrove v. Battin, 413 U.S. 149, 93 S. Ct. 2448 (1973).

Chapter 5
ORGANIZATION OF FIRE DEPARTMENTS

While the first organized efforts of fire extinguishing antedate by centuries the modern municipal fire department, it must be remembered that it has only been a relatively short time—less than one hundred years—since fire fighting was taken over by public authorities. As early as 410 A.D. fire brigades were known to the Romans who left behind them on the shores of England their "siphos" or "squirts," and these fire fighting implements were the best that England had until after the Great Fire of London in 1666.

In America in 1648 "worshipful fire wardens" were inspecting New York dwellings to see if any mud had cracked off the wooden chimneys and fireplaces, and in doing so gave rise to what was probably one of the first fire department legal problems. In order to end the indignities that these inspectors suffered from housewives who jeeringly called them "chimney sweeps," the court began to impose fines on any woman who failed to pay the inspectors proper respect.

Probably the earliest American legislation relating to firesafety was the colonial law that imposed a fine of fifteen shillings on "every person who shall suffer his house to be on fire." Another law directed each citizen to fill three of his leather buckets with water at sunset and place them on his doorstep for use in case of need.

The advent of the modern municipal fire department, with its vast resources in the technology of fire prevention, detection, extinguishment, and control, has given rise to a myriad of legal problems, and at the outset it is well to have a clear idea of the considerations involved in the creation and organization of fire departments.

The creation and composition of a fire department, whether it is under the control of a state, wholly or partially, or under the exclusive control of a local governing body, is a matter generally controlled by the local laws embodied in statute, charter, and ordinance. Although a municipal fire department is an agency of local government maintained for the benefit of the community and its composition, powers, and duties defined by charter, there also are rural fire departments organized under county or state laws. Some of these departments are autonomous political entities having their own taxing authority. As a supplement to the discussions in this book relating to all fire departments, there is included at this point material which is especially relevant to volunteer fire fighters and fire districts.

Volunteer Fire Departments

Organization

About nine out of ten fire fighters in the United States are volunteers in rural fire departments and unincorporated towns. Although many of these departments have been established under state laws authorizing the formation of fire districts, some have been organized as individual fire companies. For example, Sections 14825-27 of the California Health and Safety Code provide that fire companies in unincorporated towns may be organized by recording with the county recorder a certificate signed by the company's foreman or presiding officer and by the secretary, showing the date of the organization, the name of the company, the names of the officers, and the roll of the active and honorary members.

It is important that fire departments formed under a type of law similar to California's comply with the requirements to the letter in order for their fire apparatus to qualify as "authorized emergency vehicles" under state vehicle codes. The mere installation of a red light and siren on a fire truck does not entitle its driver to exceed the speed limits, go through stop signs, or to otherwise disregard traffic laws; such exemptions are only accorded to operators of "authorized emergency vehicles," as discussed in the next chapter. When a fire company has not been legally organized, or if its legal status has not been properly maintained, its members would not qualify for the immunity from civil liability which state laws often provide apparatus operators responding to an alarm. Members also might be deprived of the benefits which workmen's compensation laws ordinarily provide for line of duty injuries. That such a situation could exist is not purely hypothetical. At a meeting of the California State Board of Fire Services in July, 1975, it was reported that at least thirty-two "organized fire companies" and eighty-one volunteer fire departments throughout the state have a questionable legal status to perform fire services because of possible noncompliance with the state health and safety code provisions mentioned at the beginning of this discussion. Every fire department in every state should check to make sure that it has complied with the applicable laws for its formation and continuation so as to give its members all possible protection, immunities, and benefits provided under the laws of its jurisdiction.

Exemptions, Benefits, Powers, Duties

State laws may provide that fire fighters duly enrolled in authorized fire companies are exempt from jury duty and from serving in the state militia, and from paying certain taxes or fees, such as poll taxes and toll road fees. In general, volunteer fire fighters have the same rights, duties, and powers as paid fire fighters, provided that their names are duly enrolled on the fire department register, recorded as required by law, and re-recorded at required intervals.

Included are the rights to enter upon property to gain access to fire, to make rescues of endangered persons, and to generally carry out the duties imposed by their department rules.

Volunteers whose names have been properly certified can qualify for workmen's compensation benefits, as can their dependents, in the event of injuries received while acting within the scope of their employment. The fact that a person is not registered with a volunteer company will not preclude him from obtaining workmen's compensation benefits in the event of an injury in states which have adopted a law similar to that in California which provides that *anyone* who volunteers to help suppress a fire when requested by an officer or employee of a public agency is deemed an employee of that agency and is entitled to compensation.

Although workmen's compensation laws do not usually exclude coverage for high school students enrolled as junior fire fighters as long as they are trained and treated the same as the regular members of the fire department, city attorneys and county counsels usually recommend that they be required to obtain their parents' written consent to join a junior organization and to participate in its activities.

Volunteers must be provided with all the safeguards, such as protective clothing, self-contained breathing apparatus, and other safety measures, as are required by state and federal laws for paid fire fighters. Included within these safeguards is effective fire fighter training.

A fire department having authority to operate an ambulance frequently runs into conflict with a private ambulance operator who desires to be notified first in the event of a vehicle collision where injured persons require attention or transportation to a hospital. Sometimes the state highway patrol is given the authority by law to select which service should respond, in which case, the general policy may be to notify the closest ambulance service to the scene of the emergency. At other times an agreement may be reached between the fire department and the private operator that if the situation does not require immediate transportation of the injured person, the fire department will wait until the private ambulance arrives to do the transporting.

Laws have been proposed that in situations where both public and private ambulance services arrive at the scene at about the same time, the personnel who have had the most training would be deemed in charge. With state laws increasingly requiring all fire fighters to complete first aid courses and all ambulance personnel to complete prescribed emergency medical technician (EMT) training, and with many departments now providing paramedical services, it would seem logical that the best trained persons, whether public or privately employed, be placed in charge of an emergency. However, if the patient insists that he be cared for by fire department personnel, regardless of any memorandum of understanding which may exist with the private service, his wishes should be complied with. As a taxpayer he may have helped purchase the public ambulance, and from his prior knowledge and experience he may have more confidence in the fire fighter than in the private operator. He is entitled to exercise his choice.

A Volunteer's Liability

Just because a volunteer serves without pay does not render him immune from liability for gross, willful, or wanton misconduct. Even though he has been called from home to respond to an emergency in his private car does not entitle him to disregard speed laws and traffic signs on the way to the station or to a fire. Unless a state has made special provision whereby a volunteer can install a warning light, siren, sign, etc., on his own car, and then grant him certain privileges in emergency response, he must obey all the rules of the road the same as ordinary drivers. Obviously, if he has been imbibing alcoholic beverages, whether at a social event or at home, he should not drive his own car, and especially not an emergency vehicle. Any accident resulting from his inebriated condition could not only imperil his own life and that of his co-workers, but could render him personally liable for the consequences. Without a special rider on his policy, his insurance would probably not cover the accident.

It has been known to happen that a fire fighter has run down another fire fighter who belongs to his own company in responding to a call. Although the state laws usually afford workmen's compensation to the injured party working at the scene of an emergency, he may decide to sue the person who injured him anyway.

A case in point is where a volunteer fire fighter of a fire district in New York state was injured while directing traffic in front of the fire station during a fire, by being struck by the auto of another volunteer. The court held that the Volunteer Fireman's Benefit Law was his exclusive remedy and dismissed the suit. However, in another instance where a fire buff, who was also a volunteer fire fighter in New Jersey, fell through an air shaft while assisting New York City fire fighters at a fire, the city was held liable for his injuries. The decision held that members of the department knew he was untrained, yet they allowed him to accompany them to fires and to take part in fire fighting activities. Since he was not an employee of the city he was not eligible for workmen's compensation, nor could it be said that he assumed the risk or was barred from recovery on account of contributory negligence, though the city raised these defenses.[1]

Liability involved in handling rescue cases and in rendering first aid treatment is discussed later in Chapter 13, and the liability potential in driving fire apparatus is presented in considerable detail in Chapter 6. In any case, the standard of caution for a paid driver and a volunteer driver is the same; that is, they both are expected to drive a fire truck in the same way any prudent fire truck driver would drive it. The public has the right to expect that no one will be permitted to drive fire apparatus who has not been thoroughly trained and checked out as competent to operate it.

Insurance

Because volunteer (and other) fire fighters have been known to serve alcoholic beverages at some of their get-togethers, it is advisable that volunteer

---

[1] Wolf v. City of New York, 365 N.Y.S.2d 205, 47 A.D.2d 152 (1975).

associations obtain insurance to cover any liability that might arise from a motor vehicle accident involving negligence of an intoxicated member who has just left a party sponsored by the association where the beverage was served, even on private premises. Where alcoholic beverages have been served on fire department property, whether or not in violation of department rules but with the acquiescence of town officials or district directors, it is not unlikely that the latter might be joined as defendants with the fire fighters' association which sponsored the affair. With juries as unpredictable as they are, a judgment might be rendered against all the defendants. To help avoid town or fire district liability in such cases, no official sanction should be given the event, and it should be held at some place other than the fire department's facilities.

Fire Districts

In addition to the fire protection afforded by the "organized fire companies," discussed previously, as well as by municipal and county fire departments, state and federal forestry fire departments, fire services can be furnished by water districts, community service districts, and fire districts under the laws of some states. For example, there are over 400 fire districts in California.

Organization of Fire Districts

The applicable laws under which a fire district is organized bear close scrutiny. An example of the importance of following the statutory prescriptions exactly is an Illinois case where a fire protection district was held to be "fatally defective" owing to the inclusion of an island in the Mississippi River which was actually located in the State of Missouri.[2] Where a Colorado statute provided the authorization, it was held that real property annexed to the city may be excluded from the fire district.[3] A Massachusetts case held that the establishment of a fire district covering several towns is a state affair, and could become operative without a city's acceptance.[4] A city, or part of a city, may be included in a fire protection district, though some statutes forbid this unless the city is annexed as part of the district. In Illinois, where a city and a fire district have overlapping boundaries, the city must cease to perform fire protection services and stop levying taxes for this purpose, and let the fire district render the services in the duplicate area.[5] In some states fire districts are established as corporations; they are considered public corporations in New York, but not in Washington, while in Rhode Island they are held to be quasi-corporations.

Some provisions of the fire district organizational laws relate to the terms of district officers, their compensation, and what constitutes a quorum; the methods by which the board can take action (ordinance, resolution, or motion);

---
[2] People *ex rel* Curtain v. Heizer, *et al*, 36 Ill.2d 438, 223 N.E.2d 128 (1967).
[3] Dittola v. Guippe, 402 P.2d 938 (Col. Sp. Ct. 1965).
[4] Comm. v. Badger, 243 Mass. 137, 137 N.E. 261, 263 (1922).
[5] People v. Edge O Town Motel Corp., 26 Ill.2d 83, 186 N.E.2d 33 (1962).

and how it can meet its financial needs, including the procedure for obtaining additional revenue to meet expenses, such as increased costs mandated by state regulations and higher premiums charged for workmen's compensation insurance.

Powers of the Board of Directors

All the authority given to a fire district can only be exercised by its board of directors; an individual member has no power whatsoever. Without authorization by his board in a public meeting, he cannot, for example, give orders to the fire chief, or purchase an item for the fire department and then seek reimbursement for the amount he has expended.

Many fire protection districts are given very broad authority. In California, for example, it includes the right of eminent domain; to sue and be sued; to acquire real property; to employ legal counsel; to establish a water system (mains, hydrants, storage, etc.) for fire protection needs, either by contract or by acquisition or construction; and to operate not only fire equipment, but ambulances, rescue and first aid vehicles, which it can either own or lease. It also includes the right to enter into mutual aid contracts (and authorize response outside the district), establish a civil service system, provide its members with insurance to cover accidental death or injury (in addition to workmen's compensation) as well as group hospital services, either directly, by contract, or by insurance.

A fire district board in California may also maintain membership in local, state, and national fire protection organizations in behalf of its fire department and pay reasonable expenses, including travel of any of its members or employees whom it has authorized to attend professional or vocational meetings. The board may also clear land of brush, grass, rubbish, or other flammables, either by order to the owner, by contract with lot cleaners, or with its own personnel, and obtain reimbursement for any expense involved through procedures prescribed by the county auditor-controller.[6]

Board Employees

Although a board member may have been elected by the other members to be its official secretary, it is advisable to employ an executive secretary to take the minutes, keep the records, prepare the correspondence, post the notices, mail out the agenda prior to the board meeting, and in the smaller districts, draw up the warrants for payment of bills and render the monthly financial statement. In large districts an accountant is desirable in addition to the secretary.

Rules

Even though the volunteers in a fire district may have their own association and bylaws, its bylaws are not the rules of the district's board of directors. The board is the head of the fire department and the fire chief is its admin-

---

[6] Health and Safety Code of California, Part 2.7.

istrator, and it behooves a well-run district to have its own rules in codified form available to all fire fighters in its department.

While volunteers may set their own standards for membership in their association, again these are not binding upon the district's board in making appointments. While the board may (and should) confer with the officers or committee of the volunteer's association before appointing the chief, making promotions, filling vacancies in the ranks, setting salary schedules, working shift schedules, terminating members, setting policies on leaves, vacations, type and frequency of drills, uniforms, apparatus, etc., it has the sole authority for making the final decision in these matters.

For example, in one fire district the volunteer's association voted by a slim majority to admit a number of women into its membership, and its spokesman at the district's board meeting recommended that the board appoint these women to fill the existing vacancies in the volunteer's ranks. Because some of the women were wives or girlfriends of some of the volunteers, a majority of the board voted to deny them membership in the fire department, and thus avoided resignations from nonapproving members in the volunteer's association who strenuously objected not to women in general, but to these specific ones becoming members.

Records

Good records are essential to avoid legal pitfalls that might arise at later dates. For example, there should be medical records in the personnel file for every member, starting with the entrance exam and the annual checkup provided by the district thereafter. Even student fire fighters should also be given a medical exam to determine that they have not suffered football injuries, etc., that might later be confused with a compensable injury sustained at a fire. Without good medical records there is no way of knowing whether the cause of a disability was preexisting or not, and litigation could result.

Records for fire apparatus and equipment maintenance are also important. The failure to make a daily check of the air or oxygen supply in breathing apparatus and resuscitators might be very relevant in a claim against the district should a fatality occur because of an empty cylinder that no one had replaced following use at a fire or at the end of a rescue run. Driver certification records for operators of emergency vehicles, as well as certificates in emergency medical training (EMT), cardiopulmonary resuscitation (CPR), and first aid, could be very useful in defense against lawsuits alleging negligence on the part of the board in failing to see that equipment operators were properly trained, and on the part of drivers and members administering emergency assistance on rescue calls, assuming that such assistance was being carried out as taught.

Public Meetings

There are state laws regulating the degree to which public business must be conducted openly. Some of them contain exceptions to the full open meeting concept, e.g., matters relating to the security of public buildings; personnel matters, such as appointments, promotions, or dismissal of employees or officers; and conferences with employee representatives or legal counsel. Such laws may also forbid holding meetings in any building that prohibits access to women and minorities, such as a private men's club, and may also provide that you cannot require a member of the public attending the meeting to register his name or fill out a questionnaire, or do anything else as a condition of attendance. The laws also may not prohibit holding special meetings, providing that reasonable notice (usually 24 hours) has been given members of the board (or a waiver of notice obtained), and that news media have been notified. Special meetings, however, cannot be used to consider any subject other than that announced in the notice.

Fire district board members should not conduct business informally in the local cafe, at a member's home, or even in the office of the board's counsel. Suppose that one or two members go to see the board's attorney, whether it be the assistant district attorney, county counsel, or a private lawyer, and as a result of advice obtained, immediately take action, e.g., lease a truck, sell a refrigerator, etc. The action would be illegal. Instead, they should report to the full board what they have been advised, and any action taken should only be after a majority of the board has decided what needs to be done.

Executive Sessions

If an executive session is held to discuss personnel matters, for example, either the meeting room can be cleared of visitors or else the board can retire to a separate room. The fire chief, or employee representative, or legal counsel, may be present if needed to confer on the problem, but it might be well to exclude any secretarial employees. The board member who holds the title of secretary to the board can make any notes if deemed desirable, though these need not become a part of the official minutes which must be made accessible to the public; only the minutes of the public session must be kept, and upon reasonable notice, any member of the public may look at them or copy them. Votes are not taken in executive sessions; they are taken when the board reconvenes in open session. Thus, minutes of the public meeting should show the time the board went into an executive session and the reason that the public meeting was reconvened and called to order by the chairman of the board, and specify the action taken in public session on the matters discussed in the executive session.

Fire Hazard Removal Programs

Any fire department which attempts to carry out a lot cleaning program is likely to have legal problems, not only in determining the names of the owners to whom the notices should be sent but in other matters as well. If the owner fails to have the work done within the period of time specified, then the fire district may hire a contractor to do it and either have him bill the owner directly, or pay him out of district funds and have the costs added to the tax assessment for the land involved, in those states having laws which authorize this procedure. In any case, someone from the fire department should supervise the program to see that the right parties receive the notices, that only parcels which create a definite fire hazard are posted, that the same standards are applied throughout the district, that the contractor bills the right property owners and does not bill owners of lots which he has not cleaned, that he does not cut down trees and haul off valuable firewood for which the owner will later demand payment from the fire district, and that if controlled burning is attempted, either by the owner, contractor, or fire department, all the conditions of the burning permit are met, standby fire apparatus is on the scene manned and ready to go into action, and every precaution taken to prevent the fire from negligently escaping to damage adjacent premises with resulting liability.

Legal Counsel

A fire district needs good legal counsel to protect its interests in event of litigation. It may be the county's counsel or district attorney who has sufficient staff assistance available with sufficient time to interview witnesses, take depositions, prepare the briefs, and otherwise make a thorough preparation of defense in a suit. If there is some question about availability of public counsel, it might be best for the fire district to retain a private attorney to handle the case. It is not uncommon for a county to bill a district for the services of its legal staff, and retention of private counsel might bring more immediate attention to the solution of the problem, as well as avoid casting a further burden on a county's legal personnel.

Whoever serves as the district's counsel should be supplied with records, minutes, names of witnesses, photos, and all data relevant to the suit. For example, if the suit is based upon some action that the board was alleged to have taken, such as entering into a contract with an individual for the performance of certain services, and if that contract was discussed in executive session of your board, then you should be able to produce minutes of that board meeting showing that the board reconvened after the executive session and voted on the matter in open public session. If no such minutes exist, then it might be erroneously assumed that the district would not be liable on the contract.

A suit was instituted against a California fire district in which the conditions outlined in the preceding paragraph were present. The plaintiff alleged that he was granted a two-year contract to supervise the fire chief. Notwithstanding a transcript of a tape recording of the executive session that showed that the board members did not assent to such a contract, and despite the fact that the minutes of the board meeting failed to show that it ever reconvened in public session at all, or voted on the contract, and even though two out of the four board members present at the meeting later signed affidavits that they had never voted to enter the contract in an open meeting, and even though a dozen reputable persons, including the fire chief, signed affidavits that they were present at the meeting the entire time and knew that the board never did reconvene in a public session after its executive session, nor did it ever publicly vote to enter the contract, and even though the minutes of the meeting in question were corrected at the next board meeting, on the motion of one of the members who was alleged to have agreed to the contract, to show that no such agreement had ever been reached in either the executive or public session, the superior court ruled that the fire district had to pay the plaintiff two years salary for work that he had never performed. The court reasoned that parol (word of mouth) evidence was admissible from the former board chairman to dispute the minutes and other evidence; also that there was a rebuttable presumption that a public board acts lawfully, and that even if it had not, it would not affect the rights of the plaintiff who was not responsible for the conduct of the board members.

Director's Liability

Directors are liable the same as private individuals for their own negligent and wrongful acts. They are also liable for the wrongful expenditure of public funds. They cannot make gifts of public funds or department property, though the law sometimes provides that a transfer of government property may be made by one government agency to another; e.g., to the county's civil defense agency, without a transfer of funds. Directors cannot make private use of public property, such as borrowing the fire department's pickup truck for use on personal projects.

Although the board may use district funds to print a fact sheet on a bond issue, for example, it is illegal to use public funds for publicizing candidates or financing political activity.

Directors, as public officers, are not personally liable for the acts of members of the fire department unless, as pointed out in the discussion on liability of superior officers (Chapter 12), they instigate or direct the acts, participate in them, ratify them, or acquiesce in them.

A discussion in *Fire Department Personnel Reporter*[7] of the liability of fire and police commissioners who have been sued as the result of damages sustained by a discharged employee cited an Illinois case in which the commissioners were held to be immune from liability if they acted with good faith and within the scope of their authority.[8] The discussion also distinguished the lack of immunity if a board member knew or reasonably should have known that the action which he took within the sphere of his official responsibility would violate the constitutional rights of the person affected, or if he took the action with the malicious intention of causing a deprivation of his constitutional rights: the latter situation involved a school board and a student, and the Supreme Court of the United States affirmed a 1973 8th Circuit Court decision.[9]

Both criminal and civil liability can result for accepting bribes, favors, gratuities, or "kickbacks" on the purchase of fire equipment, for example, or for selling property to the district at an inflated value, either directly or through "dummy" parties, or otherwise violating the laws relating to conflict of interest.

---

[7] FIRE DEPT. PERS. REP., May 1975, p. 6.
[8] Reich v. City of Freeport, 388 F.Supp. 953 (N.D.Ill. 1974).
[9] Strickland v. Inlow, 485 F.2d 186 (8th Cir. 1973), *aff'd* Wood v. Strickland, 95 S. Ct. 992 (1975).

Conflict of Interest

Many states have laws that deal with conflict of personal or financial interest of public officials. In Terry v. Bender,[10] a California appellate court decided in 1956 that a mayor's pecuniary advantage in the continuation of a contract between the city and a special attorney employed by the city tainted the contract with illegality and rendered the payment to the attorney contrary to the public interest. The court opined that "a public office is a public trust created in the interest and for the benefit of the people. Public officers are obligated, *virtute officii,* to discharge their responsibilities with integrity and fidelity. Since the officers of a governmental body are trustees of the public weal, they may not exploit or prostitute their official position for their private benefits. When public officials are influenced in the performance of their public duties by base and improper considerations of personal advantage, they violate their oath of office and vitiate the trust reposed in them, and the public is injured by being deprived of their loyal and honest services. It is therefore the general policy of this state that public officers shall not have a personal interest in any contract made in their official capacity . . . A transaction in which the prohibited interest of a public officer appears is held void both as repugnant to the public policy expressed in the statutes and because the interest of the officer interferes with the unfettered discharge of his duty to the public. The public officer's interest need not be a direct one, since the purpose of the statutes is also to remove all indirect influence of an interested officer as well as to discourage deliberate dishonesty. Statutes prohibiting such 'conflict of interest' by a public officer should be strictly enforced."

Other Considerations

In addition to the applicable procedural pointers presented in the final chapter of this book, there are a number of suggestions that pertain especially to directors of fire districts that might assist in solving or avoiding some legal problems, including:

1. *Dual Membership in the same Department:* It is difficult for a person to be a volunteer fire fighter who must take orders from the fire chief on drill nights and at fires, and concurrently be a member of the board of directors which holds the fire chief accountable for the efficiency of the drills and the handling of the fire. To avoid criticism of fellow board members for being a "pipeline" to the volunteers, it would be advisable for a director to either resign from the department or take a leave from the ranks upon being elected to public office.

2. *Keeping File Copies of Important Data:* It is good practice to retain copies of important departmental papers in a secure but available manner. If a board meets in a room where there is no ready access to permanent department files, then it might be well to establish a portable file case containing the current year's issues of board meeting agendas, minutes, financial reports, department's budget, fire chief's annual report, board's latest inspection report,

---

[10] Terry v. Bender, 143 C.A.2d 193, 300 P.2d 119 (1956).

latest Insurance Services Organization (or other rating organizations) report on the department, mutual aid agreements, opinion given by your legal counsel on questions submitted by the board, latest amended copy of the fire department rules adopted by the board, and a file of board resolutions covering policy matters having continuing application.

3. *Public Information:* Conflicting information to the news media and the public must be avoided. Policy should be to refer inquiries to the fire chief in regard to fire department activities, and to the board chairman on matters relating to policy decisions. It should not be necessary for the officers of the department to call a board member to determine whether or not he should use department apparatus to remove a cat from a tree, replace a flag pole halyard, pump out a flooded cellar, or other nonfire protection requests. Such policies should have already been established by the board and promulgated to the members.

4. *Promotional Examinations:* Although this subject is discussed in more detail in a later section, a word of caution is in order at this point. It is better for the board to arrange to have promotional examinations administered by professionally qualified persons who are not associated with the fire department; to do otherwise could open the door to charges of partiality at the very least. The extent of the board's participation should be to interview the candidates who pass the written test for the position of fire chief, and, other factors being equal, select the man who heads the list. Promotions in the lower ranks can usually be left to the discretion of the chief. In departments having assistant and battalion chiefs, the board may wish to confer with the fire chief in making such appointments based upon such essential qualifications for promotion as knowledge, experience, and merit.

5. *Discipline:* The board's rules for the fire department should provide that the fire chief be given the discretion to suspend any member for serious infraction of the rules, and to impose lesser penalties for minor infractions. Provisions should also be made for a grievance procedure and hearing at the request of a member who has been suspended or recommended for dismissal. In smaller departments this hearing could be conducted by the board. Court decisions affecting these matters are discussed in the next chapter.

Police-Fire Combination Departments

Although there have been a number of instances where dual functions of fire and police protection have been assigned to public safety departments, in the absence of a statute prohibiting such an arrangement there should be no legal objection as these functions are usually considered a matter for local determination. Under the applicable statutes of Illinois, for example, the City of Peoria had the power to adopt an ordinance combining to some extent the duties of the fire and police departments.[11] But notwithstanding the fact that a

---

[11] Hunt v. City of Peoria, 30 Ill.2d 230, 195 N.E.2d 719 (1964).

combined police and fire department has been in operation in Oakwood, Ohio, for a number of years, a suit was filed contesting the constitutionality of the adoption of ordinances in Blue Ash, Ohio, which established a new Public Safety Department and combined the roles of fire fighters and policemen. Because Ohio had a statute providing that fire departments "shall protect lives and property of the people in case of fire . . . and such other duties as are provided by ordinance," the supreme court of that state held that a city ordinance requiring the fire department to operate the police radio was in conflict with the general law, for "such other duties" must mean duties incident to one's position.[12] A 1975 decision by the Supreme Court of Vermont held that the City of Montpelier could not compel fire fighters and policemen to be represented by the same union, even though the city claimed that it would save time and would be more expedient to negotiate both contracts at one time, with one bargaining representative.[13]

Are Fire Fighters "Public Officers" or "Employees?"

While there is no distinction in law between a "member" of a department and an "employee" of a department, quite a controversy exists as to whether a fire fighter is an "officer" or "employee." Fire fighters may be classified as either officers or employees. Because fire fighters are often vested with powers of inspection and the enforcement of laws relating to fire prevention, as well as duties relating to fire suppression and investigation, they have been considered as officers. The fire fighter is also subject to detailed supervision of supervisors; his salary is for services rendered, and in most jurisdictions is not paid as an incident to an office.

In each instance reference must be made to the charter or statute governing the service, as well as to judicial decisions construing such provisions, to determine the question. With reference to paid fire departments, while there are some decisions to the contrary, and some statutes or charter provisions which expressly designate fire fighters as public officers, it is held in many cases that fire fighters and officials of the fire department are not public officers. In the same state, the courts may differ. Thus, in California, on one occasion the position of fire fighter was declared to be a public officer while on a later occasion the court, without specifically overruling the earlier case, held that a fire fighter is not a public officer. In New York City fire fighters are held not employees within the labor laws. (The chief counsel for the fire fighters' association in the state of New York, in 1950, declared: "I am firmly of the opinion that the chief engineer of a fire department, paid or volunteer, is a public officer within the scope of the court decisions affecting the issue. Indeed, there is direct judicial authority on the point in State v. Jennings, 415,

---

[12] Vair v. City of Ravenna, 29 Ohio St.2d 135, 279 N.E.2d 884 (1972).
[13] IAFF Local 2287 v. City of Montpelier, 133 Vt. 175, 332 A.2d 795 (1975).

where it was said: 'So that the real question in the case is whether the fireman is an officer, or in this case, whether the firemen for whose employment provision is made in the ordinance of 1897 are officers: for that a position in the fire department of a city may have such duties attached to it as to constitute an office is not questioned. The chief of a fire department performs such duties as make him an officer.[14']")

The importance of the distinction between "employee" and "public officer" has been the subject of much discussion, and is well illustrated in subsequent portions of this book dealing with a fire fighter's power and authority, his removal from office, the abolishing of his position, the reduction or assignment of his salary, as well as the effect the distinction has on workmen's compensation rights, municipal liability, the duty owed by the public to fire fighters, a fire inspector's liability, and the duty owed by fire fighters.

Are Fire Fighters "Peace Officers?"

Because fire departments are often responsible for detecting fire code violations and for apprehending arsonists, the question has arisen whether fire fighters are "peace officers." Do they have the authority to carry concealed weapons? Are they required to complete training courses in the handling of firearms and learn the application of the laws of search and seizure in those states where all "peace officers" must take such training?

There are code provisions in some states which specifically designate fire fighters as peace officers. A recent opinion of the California Attorney General concluded that only those fire department personnel whose "exclusive duty" is the enforcement of laws relating to fire suppression and prevention qualify under the provisions of Chapter 4.5 of the California Penal Code as "peace officers." It would appear that under this interpretation, fire department members assigned to the arson investigation unit could qualify as "peace officers," but members assigned to the fire prevention bureau who might also be called upon to fight fires in a major emergency would not qualify. Neither would the members of most fire departments, regardless of assignment, because of departmental rules requiring them to report for fire suppression duty at any time there is a "recall" ordered as a result of a large scale emergency.

Although the above California opinion referred to that state's penal code, a similar view would probably be applicable to the section of the state's Health & Safety Code (13875) which provides that officers of fire district fire departments should have the powers of "peace officers" while engaged in the performance of their duties with respect to the prevention and suppression of fires and the protection and preservation of life and property against the hazards of fire and conflagration.

In any event, all fire fighters whose duties include the enforcement of codes and ordinances, irrespective of their qualification as peace officers, should

---

[14] 57 Ohio St. 415, 49 N.E. 404 (1898).

become familiar with the proper procedure for making arrests, making searches, seizing evidence, etc. Furthermore, no member should carry a concealed weapon until he is positive that he is in compliance with the laws of his jurisdiction regarding such possession and has had thorough training in its use and handling.

Federal Impact on Fire Departments

Insofar as management controls were concerned, there was a time when a fire chief had mainly to consider his local charter and ordinance provisions; then the state legislature became a power to reckon with. Today, however, the U.S. Congress is beginning to exert increasing influence.

Acronyms for some of the federal acts and agencies which affect the fire service include NFPCA (National Fire Prevention and Control Administration), DOT (Department of Transportation), FLSA (Fair Labor Standards Act of 1938), NBS (National Bureau of Standards), HEW (Department of Health, Education, and Welfare), OSHA (Occupational Safety and Health Administration), DCPA (Defense Civil Preparedness Agency), HUD (Department of Housing and Urban Development), and SLFAA (State and Local Fiscal Assistance Act of 1972). Of these, the creation of the NFPCA will probably have the most favorable long-run benefits for the fire service, while the FLSA, if ruled to be applicable to fire departments, is most likely to result in divergent interpretations and legal disputes. Some of the areas in which federal regulation affects the fire service are discussed below.

Manning Practices

The Fair Labor Standards Act of 1938 was amended in 1974 to include public employees engaged in fire suppression activities. The changes became effective January 1, 1975. However, action by Justice Burger of the U.S. Supreme Court suspended application of the final revision of the regulations applying to fire and police personnel.[15] Regardless of the final decision, many fire departments throughout the nation have renegotiated their contracts to reflect the FLSA changes in maximum hours, time and one-half for overtime, etc. It would seem advisable to comply with the new regulations pending the final decision, for if the Supreme Court affirms the circuit court opinion upholding the constitutionality of the regulations, it would be retroactive in application to the day they became effective.

Exempted from the operation of the FLSA are fire departments with less than five paid members. Volunteers and fire commissioners or directors who receive no fees (which are considered salary) would not be counted in determining this number. Though the regulations might apply to a fire department because it has five or more paid men, supervisory personnel who have the authority to recommend the hiring and firing of members and who otherwise

---

[15] National League of Cities v. Brennan, 95 S. Ct. 532 (1974), interim order granting injunction, 31 Dec. 1974; as of March 1, 1976 no further decisions have appeared (through 96 S. Ct. 890).

qualify by reason of responsibility and high salary are exempt from its provisions. While it is doubtful that a fire fighter will qualify for an exemption merely because he is designated as supervisor of a piece of fire apparatus, there is little question that students hired to sleep at a fire station during the brush fire season will come within the regulations.

Safe Working Conditions

The Occupational Safety and Health Act (OSHA) was signed into law on December 29, 1970, and became effective on April 28, 1971. Employers covered by this act are required to provide safe and healthy working conditions and to comply with applicable health and safety standards. Paid fire departments, if part of a political subdivision, are required by the law to comply with OSHA standards as administered by state agencies as a condition for the state's receiving federal help to enforce the law.

Even if not a political subdivision; i.e., not organized under some statutory authority, a fire department will be covered under OSHA if there is an employer-employee relationship. This would be the case in private fire departments or volunteer departments having paid employees or where members receive pay for drills or calls. Even though no one in a fire department or on its board of directors receives pay or fees of any kind, the fire station might still be considered a "place of employment" by the state agency that is applying the OSHA standards, and its compliance officers will probably require compliance with applicable safety standards the same as in paid fire departments.

Insofar as many fire departments have been inspected for compliance with OSHA standards and have had penalties imposed for violations, it would be well to obtain OSHA standards and comply with them as soon as possible. One of the most common violations has been the failure to keep a record of all injuries and fatalities on prescribed forms (obtainable from regional OSHA offices). Pilot inspections have disclosed other violations such as a lack of first aid kits, insufficient exit doors, doors that swing the wrong way out of assembly rooms, lack of signs for exits and for fire extinguishers, lack of "No Smoking" signs (in battery recharging area), lack of labels on electric panels, undergrounded electric tools, unshielded grindstones, lack of handles on files, lack of inspection tags on fire extinguishers, etc.

Problems in enforcement of fire safety ordinances can arise if a fire marshal gives a property owner one set of orders and a federal or state compliance officer gives him a conflicting set of orders. To avoid such a possibility, both enforcement officials should apply the same set of standards, or, at the very least, make joint inspections where feasible. In that way, each inspector will know what the other is trying to accomplish, thus permitting them to seek a mutually satisfactory solution to the problem.

Dispatching

Following the passage of the federal Emergency Medical Systems Act (EMSA) in 1973, which requires that emergency medical services systems shall participate in a central communications system using the universal telephone number 911, members of the medical profession have become increasingly interested in the expansion of the 911 systems. The Federal Communications Commission (FCC) does not advocate the removal of street fire alarm boxes, though this has been done in such cities as Los Angeles, but it does believe that 911 will improve emergency communications. Over 400 of the 911 systems were installed in the United States as of January 1, 1974, and some states are adopting deadlines for state-wide installation; e.g., a California law requires it by 1982. It is the Secretary of the Department of Health, Education and Welfare who determines the period of time within which the emergency medical services must utilize the 911 system.

Training

Funds for conducting training courses for emergency medical technicians (EMTs), as well as for buying ambulances, have been made available through the National Highway Traffic Safety Administration (NHTSA), which is under the Department of Transportation, by allotments to all states. The Rural Development Act, administered by the Department of Agriculture, also authorizes funds for emergency medical care training.

The plans of the National Fire Prevention and Control Administration (NFPCA) include a training program for fire officers within the framework of a national fire academy that is still (1975) in the planning stage. There should be little doubt that the fire service will receive enormous benefits from the research and training that this new federal agency hopes to provide, assuming that it receives continued congressional support in the matter of appropriations in the years ahead.

Personnel Selection

The federal Equal Employment Opportunity Commission (EEOC), through its power to accept complaints and hand down decisions on alleged discrimination in hiring practices, has been instrumental in bringing about changes in the customary criteria for selection of recruits for the fire service. One area where its effect has been felt is its stance regarding entrance examinations. For example, an entrance exam that 42 percent of black applicants failed, but only 29 percent of whites failed, was held to violate the act as not being job related and for creating a "disparate effect." The EEOC also has ruled that a requirement that applicants who have served in the armed forces must have an honorable discharge is not a valid prerequisite, for twice as many blacks receive dishonorable discharges as whites, indicating "racism" as the most significant factor in this disparation.

The commission has also ruled that arrest records cannot be used to disqualify applicants, as experience has shown blacks are arrested substantially more frequently than whites in proportion to their numbers. And unless it can be shown that facial hair needs to be regulated as a business necessity, it is discriminatory to bar applicants with beards, mustaches, and goatees, since these are much more prevalent among black males than among whites, for these are sometimes a symbol of black racial pride and cultural heritage.

The EEOC also ruled in March of 1975 that a minimum height requirement of 5 ft. 7 in. violates the act because it disproportionately disqualifies females and Spanish-surnamed Americans.

Although the EEOC had adopted guidelines on employee selection procedures, the full impact of the guidelines on the fire service was not felt until 1971 when the Supreme Court of the United States adopted these guidelines in determining job-relatedness of entrance examinations in Griggs v. Duke Power Company.[16] The court held that a disparate effect in hiring practices can only be justified if the examination is job related.

Once a complainant has established a *prima facie* claim of discrimination, the burdens fall upon the employer to show that the discriminatory effect of the questioned examinations is based upon job performance standards. Tests must be relevant to the job requirements and must be validated by such techniques as the requirement for fire fighters to take a battery of tests and then the comparison of their scores with the department's evaluation of their work; another method is to put the test data away for a year and at a later date comparing it with the job performance ratings.

Unless a logical relationship can be shown between the test material and what a person will do on the job for which he is being tested, the test lacks content validity. To be completely certain that a test is valid, and to be on the safe side, tests should not be used simply because they have been validated for another department. Rather, the tests must be applicable to the department in which they are conducted.

Having a job related test is not enough to satisfy the Equal Employment Opportunity Act; it must also be properly administered. On June 25, 1975, the United States Supreme Court found that the tests in question were improperly administered because they were given selectively and were not validated on younger, inexperienced, nonwhite applicants, but on experienced white workers; moreover, the job performance ratings given by the supervisors to validate that the exams were not sufficiently objective to identify the specific criteria employed by them in making the ratings.[17]

Other court decisions have modified the traditional selection criteria; for example, in an early Minnesota case, the court ordered the civil service commission to give absolute job preference to 20 minority applicants on the next examination for Minneapolis fire fighter. Further, it ordered the commission to lower the minimum age from 20 to 18, raise the maximum age from 30 to 35 (at least until the 20 minority applicants were hired), remove the requirement of a high school diploma or a G.E.D. equivalency certificate, delete any reference to arrest records, accept an applicant with a felony or misdemeanor record if the felony is more than 5 years old or the misdemeanor more than 2 years old, and, finally, disregard the time interval between conviction and application for employment entirely if the crimes could not be proved to have a bearing on the applicant's fitness as a fire fighter.[18]

---

[16] Griggs v. Duke Power Company, 401 U.S. 424, 91 S. Ct. 849 (1971).
[17] Albemarle Paper Co. v. Moody, ___ U.S. ___, 95 S. Ct. 2362 (1975).
[18] Carter v. Gallagher, 452 F.2d 315 (1972).

Background Investigations

Regulations restricting the dissemination of criminal history information have been issued by an agency of the United States Department of Justice known as the Law Enforcement Assistance Administration (LEAA). They became effective on June 19, 1975. Although criminal justice agencies can be given such information for legitimate purposes, including pre-employment checks, it is unlawful to pass it along to anyone else except under specific conditions. Therefore it would be futile to request the police department to run an FBI check on an applicant's fingerprints if it is illegal for the police chief to tell the fire chief what the check disclosed.

However, the regulations do permit the disclosure of the arrest of an employed fire fighter under certain conditions, and they also provide that other public agencies may be given the records of arrests (up to one year back) and records of convictions, if permitted by state statute, executive order of the governor, or by court rule as a condition of employment.[19]

As indicated in the previous discussion on permissible screening methods under equal employment opportunities guidelines, the use of criminal records as a ground for disqualifying applicants is not looked upon with favor. But should a background check disclose that an applicant is wanted for parole violation in another state or is an escaped convict, or otherwise a fugitive from justice, the fire department would probably not be required to accept his application until after he had served his sentence or had otherwise paid any penalty imposed upon him.

Minority Quotas

In September, 1974, the United States Court of Appeals ruled that the Massachusetts Civil Service must certify to Boston and Springfield one minority fire fighter applicant for every white fire fighter applicant. In 138 other Massachusetts communities with minority populations exceeding 1 percent, it was required to certify one minority applicant for every three white applicants for the fire department, and it was ordered to continue this procedure until the proportion of minority fire fighters employed is the same as the proportion of minorities in the cities employing them.[20]

Following the example of the decision in the Massachusetts case[21] which the United States Supreme Court refused to review, a Federal court in 1974 required the City of Philadelphia to appoint one minority fire fighter for every two whites. And in California, the Fair Employment Practices Commission (FEPC), in furtherance of the "affirmative action" program, prevailed upon the cities of Los Angeles and San Francisco to institute various measures to increase the ratio of minority membership in their fire departments, including imposing quotas, revising entrance exams, and lowering the age requirement to 18. After all the blacks on the San Francisco eligibility list were appointed, the

---
[19] 40 *Fed. Register* 22 114, May 20, 1975.
[20] Boston Chapter NAACP, Inc. v. Beecher, 504 F.2d 1017 (1st Cir. 1974).
[21] *Ibid.*

chief had an urgent need to fill more vacancies. He then certified that public safety required the immediate appointment of more fire fighters, and the court allowed him to fill sixty vacancies from an all white list as a temporary expedient until a new list could be established using a validated exam which the court said had to be devised before January 31, 1975.

The first quota imposed upon the San Francisco Fire Department required the hiring of one minority applicant for every white until 118 minority applicants had been hired. In the event that another quota should be imposed upon the San Francisco Fire Department, the city will probably renew its appeal against such action on the grounds of reverse discrimination. Its appeal from future preferential hiring orders, made in 1975, was denied by the circuit court on the grounds that this was not in issue at the time, and would be purely speculative until such a new order is actually promulgated.

Though minority race hiring quotas have been imposed in several major cities having a large percentage of black population, use of quotas has not been universally favored by the courts. A United States district court in Pennsylvania found that even though Pittsburgh had failed to sustain the burden of proving that its 1972 fire fighter exam was job related, it refused to impose a hiring quota, indicating that a minimum quota might become a maximum quota, as part of the reasoning in the matter.[22] An appeal from that ruling was dismissed by the circuit court of appeals on November 18, 1974.

State Legislative Regulation of Local Fire Departments

Just how far a state legislature may go in regulating the membership, hours, wages, and management of local fire departments without infringing upon the rights of "home rule" for municipal government is a question that has received diverse answers in various jurisdictions. In West Virginia, a statute placed paid municipal fire departments under civil service and authorized the appointment of members of the civil service commission by chambers of commerce and local trade boards. When the mayor of Charleston refused to appoint a commissioner, he was compelled to do so by a writ of mandamus, despite his contention that the statute was unconstitutional. The court only held void that portion of the law which limited the selection of fire fighters to the "two great political parties" and upheld the remainder as being severable.

Working Hours

A state's imposition of a platoon system upon local fire departments is a controversial subject. In Indiana, a statute was held constitutional which required every city over 15,000 population having a paid fire department to divide the fire force into two platoons and not to exceed a twenty-four hour tour of duty at any one time.[23] A similar law was upheld in Ohio as applied to a

---
[22] Commonwealth of Pa. v. Glickman, 370 F.Supp. 726 (D.Pa. 1974).
[23] State v. Morris, 199 Ind. 78, 155 N.E. 198 (1927).

charter city.[24] But a Montana statute requiring cities to set up a three-platoon system was held unconstitutional.[25] While a Rhode Island court did not expressly declare a two-platoon statute invalid, it refused to grant a writ of mandamus to compel the Pawtucket City Council to pass an ordinance that would comply with the statute on the grounds that *mandamus* would not lie to compel a legislative act.[26] Where the people of Seattle, Wash., by direct vote put the marine fire force on a two-platoon basis, the Washington court held that the city council had no power to subsequently change it to a three-platoon system without a referendum of the people.[27]

An Ohio statute that regulated the working hours of fire fighters in cities was held to be an unlawful interference with municipal officers and a violation of the state constitution's home rule provision.[28] In Missouri it was held that a constitutional provision forbidding any law to be enacted which fixes the powers or compensation for a constitutional charter city makes a general statute relating to fire departments not applicable to home rule cities.[29]

A Michigan law giving fire fighters one day off every four days, and twenty days vacation was held unconstitutional as an unwarranted interference with local self-government and an invalid exercise of the police power. The court said, "It is well settled that a city's fire department is distinctly a matter which concerns the inhabitants of the city as an organized community apart from the people of the state at large, peculiarly within the field of municipal activity and local self-government." However, in Louisiana, the court held that the state legislature may require that vacations be granted with pay.[30]

In California, a statute requiring cities, fire districts, etc., having regularly organized paid fire departments to grant regular or permanent members thereof leaves of absence from active duty of not less than fifteen days in each year and four working shifts in every month of service, was construed as follows:

1. It does not affect salaries to be paid fire fighters so as to violate the constitutional provision prohibiting passage of a local or special law affecting salaries or fees of any officer.

2. It can be enforced by fire fighters in a mandamus action against a city of the fifth class organized under the municipal corporation act, such as Santa Ana (involved in a 1944 decision).

3. It does not apply to a county or a township, neither of which is a municipal corporation.

4. The statute does not require a municipal corporation to maintain a regularly organized paid fire department. It merely requires a city which has such a department organized to give the fire fighters the leaves of absence from active duty set forth in the law. This sufficiently distinguishes the Santa Ana

---

[24] Bond v. Littleton, 87 Ohio App. 183, 94 N.E.2d 398 (1949).
[25] State *ex rel.* Kern v. Arnold, 100 Mont. 346, 49 P.2d 976 (1935).
[26] McLyman v. Holt, 51 R.I. 96 P.2d 976 (1935).
[27] Stetson v. Seattle, 74 Wash. 606, 134 P. 557 (1913).
[28] State *ex rel.* Strain v. Huston, 65 Ohio App. 139, 29 N.E.2d 375 (1940).
[29] State *ex rel.* Sanders v. Cervantes, 480 S.W.2d 888 (Sup. Ct. Mo., en banc 1972), granting writ of mandamus denied in Sanders v. Cervantes, 423 S.W.2d 791 (Sup. Ct. Mo. Div. 2 1968).
[30] New Orleans Fire Fighters Assn. Local 632 v. New Orleans, 230 So.2d 326 (Ct. App. La. 4th Cir. 1970), aff'd 255 La. 557, 232 So.2d 78 (1970).

case from the City of Lodi (California) case in which it was ruled that a state law providing for the creation of a police relief, health, life insurance and pension fund was not mandatory, in that it permitted municipal corporations to take advantage of its provisions but did not require them to do so. It indicated that an unchartered city would have to follow the state law's provisions if it adopted a pension system.

5. A chartered city would not be bound by the decision made in the Santa Ana case, for the legislature has a much greater scope of power to regulate unchartered cities.

Salaries

In declaring a Montana statute unconstitutional which required cities to pay a stated minimum salary to fire fighters, the court said: "In general, in those states which have applied the theory of local self-government . . . it is held that acts of the Legislature attempting to fix the amount of compensation to be paid firemen, to direct the levy of a compulsory tax for a firemen's pension fund, and attempts to regulate the hours of employment of firemen which would not be a valid exercise of the police power, are an undue invasion of the rights of municipalities on the theory that the establishment and maintenance of a fire department is a proprietary function."[31]

In California, salaries are a municipal affair,[32] but in some states the legislature may fix the salaries of fire fighters, such as Nebraska,[33] Tennessee,[34] and Texas.[35] In Nebraska a city can not provide a salary different from that set by state law, but a different salary is permissible if the legislative intent was to merely set the minimum salary, according to the Arizona court.[36] A Pennsylvania statute was upheld that required political subdivisions to pay salary to fire fighters temporarily totally disabled in performance of duty.[37]

Pensions

In Nebraska, the legislature constitutionally required certain class cities to pension superannuated fire fighters and to pay such pensions from the funds of the fire department.[38]

A similar conclusion was reached in Texas,[39] North Carolina,[40] and Wisconsin;[41] however, pensions provided by charter are regarded as a local function in California.[42] Judicial construction of pension laws is discussed in more detail in Chapter 9.

---

[31] State v. Arnold, 100 Mont. 346, 49 P.2d 976, 100 A.L.R. 1071 (1935).
[32] Lossman v. Stockman, 6 Cal. App. 324, 44 P.2d 397 (1935) and Popper v. Broderick, 123 Cal. 456, 56 P. 53 (1899).
[33] Adams v. Omaha, 101 Neb. 690, 164 N.W. 714 (1917).
[34] Smiddy v. Memphis, 140 Tenn. 97, 203 S.W. 512 (1918).
[35] Austin Fire and Police Depts. v. Austin, Tex. Civ. App., 224 S.W.2d 337 (1949).
[36] Luhrs v. Phoenix, 52 Ariz. 438, 83 P.2d 283 (1938).
[37] Iben v. Borough of Monaca, 158 Pa. Super 46, 43 A.2d 425 (1945).
[38] State ex rel. Haberlan v. Love, 89 Neb. 149, 131 N.W. 196 (1911).
[39] See note 35.
[40] Great American Insurance Co. v. Johnson, 257 N.C. 367, 126 S.E.2d 92 (1962).
[41] Barth v. Village of Shorewood, 229 Wis. 151, 282 N.W. 89 (1938).
[42] Richards v. Wheeler, 51 P.2d 436 (1935) and Lossman v. Stockman, 44 P.2d 397 (1935); see note 32.

Residency Requirements

To some it would appear desirable that fire fighters live in the communities where they work, not merely to enable them to vote on bond issues, pension proposals, and other matters affecting their welfare, but also to make themselves more readily available in the event of a major emergency. Moreover, any increase in taxes that would inevitably follow their obtaining higher salaries, shorter working hours, better pensions, more fire trucks, new fire stations, etc., would be a burden on property owners that they would be able to share in, along with the rest of the citizens in their community.

From the standpoint of minority groups who are rapidly becoming the population majority in some of the larger cities, it would appear that civil service employment opportunities would be substantially increased if those who have moved outside the city were ineligible to take examinations for municipal jobs.

Several courts in the nation have ruled on the validity of the residency requirement. An Ohio court in 1973 found that it unconstitutionally violated an employee's Fifth Amendment freedom to travel.[43] The Michigan Supreme Court upheld a Detroit Civil Service Commission regulation requiring continuing residency for municipal employees.[44] The California Supreme Court, on November 8, 1973, upheld the validity of the City of Torrance's residency requirements. In Smith v. City of Newark,[45] the court ruled that a law passed by the New Jersey Legislature in 1972 which eliminated the residency requirement as a condition of employment or promotion of municipal fire fighters was not unconstitutional, and superseded previous residency rules, since a later law is deemed to repeal a former one where there is a conflict. As a result, the plaintiff, though a resident of Newark, was unsuccessful in obtaining the removal of thirty-five names of nonresident fire fighters, who had placed higher than he did, from the Captain's promotional eligibility list.

In order to supersede city ordinances, legal opinions, administrative orders, and court decisions imposing or sustaining the validity of residence requirements, voters can amend the state's constitution so as to prohibit such a requirement.

A state-wide election was conducted in California on November, 5, 1974, wherein a constitutional amendment was adopted as follows: "A city or county, including any chartered city or chartered county, or public district may not require that its employees be residents of such city, county, or district; except that such employees may be required to reside within a reasonable and specific distance of their place of employment or other designated location."

This type of law overcomes the objection of employees who live in another city, but still live close by the city of their employment, as would be the case of Los Angeles city employees who live in Beverly Hills (surrounded on all sides by the City of Los Angeles) or in any one of a number of cities which are surrounded by the larger city. Where an employee lives in a city covering hundreds of square miles, as in the case of Los Angeles, it is possible for a fire

---

[43] F.O.P. Youngstown Lodge v. Hunter, 303 N.E.2d 103 (Ohio Cm. Pls. 1973).
[44] Gantz v. City of Detroit, 392 Mich. 348, 220 N.W.2d 433 (1974).
[45] Smith v. City of Newark, 128 N.J. Super. 417, 320 A.2d 212 (Law Div. 1974).

fighter living inside the city, e.g., in the San Fernando Valley, to be over 40 miles from any station to which he might be assigned in the city's harbor area. On the other hand, any fire fighter who lives 40 miles from his station in Washington, D.C., might be living in Baltimore, Md. Whatever rule is adopted, in defining a "reasonable distance," it should probably refer to a geographical entity's boundaries rather than place of employment, when applied to employees who can be transferred periodically from one location to another within a very large area.

Though a residency requirement may be ruled valid by the highest court in the state, further questions may arise as to what constitutes a "resident." For example, where a Detroit fire fighter's family lived outside the city and he lived inside, he was told that he could not meet the residency requirement unless his wife and family lived in Detroit; upon his refusal to comply with an order that they be moved back into the city he was dismissed from his position. Although the circuit court held that he could not be considered a resident of the city while his family lived elsewhere, the Michigan Court of Appeals reversed the lower court's decision and held that it was error for both the civil service commission and the circuit court to rely solely on the residence of the plaintiff's family in determining his residence; other factors must be considered, for domicile is essentially a matter of intent, depending upon the facts and circumstances in each case.

Quoting from the residency requirement itself, the court also suggested that it may be "in the best interests of the city" to sometimes waive the requirement in order to retain employees whose personal hardships and difficulties keep them from complying with it. The court granted the plaintiff a new hearing and ordered that a waiver for him be an issue discussed at the hearing.[46]

A similar situation involved a City of Paterson, N.J., fire fighter who had lived with his in-laws and helped to pay the rent while his wife and children lived in their home in Hawthorne, N.J. Though the fire fighter voted and attended church in Paterson, he slept about three nights a week in his Hawthorne home. The New Jersey Superior Court held that he violated statutes requiring municipal officers to reside within the city.[47] A person may have more than one residence, but he cannot have more than one domicile. The domicile is the place of his abode where he has the present intention of remaining and to which, if absent, he intends to return.

Control of Fire Department

In the larger cities, the fire department is frequently under the control of a board of fire commissioners, and in some cities the fire and police board is one and the same board. Such a board, with certain powers and duties conferred upon it, is frequently beyond the control of the municipal council or governing legislative body as to such powers and duties.

In Indiana, a state law was declared void which created a board of metro-

---

[46] Grable v. City of Detroit, 48 Mich. App. 368, 210 N.W.2d 379 (Div. 1 1973).
[47] Morse, H. Newcomb, "Legal Insight: Residency Requirements," *Fire Command!*, Vol. 39, No 9, Sept. 1972, pp. 26, 27.

politan police and fire departments having members appointed by the legislature[48] but a similar act was sustained in Nebraska, where it was held proper to give the governor the power to appoint fire and police commissioners of metropolitan class cities.[49]

Federal legislation and court decisions exert more and more control on selection, appointment, and promotion of fire fighters, as discussed elsewhere in the chapter. State legislative bodies also have become involved in such matters as working hours, salaries, vacations, leaves, pensions, etc. Insofar as these same subjects are also being negotiated in collective bargaining agreements between representatives of labor unions and a city's administrative offices, it would seem that the fire board, fire commissioner, or other head of the fire department is losing some of the control over the affairs of his department.

Attempts, too, are being made to limit the prerogative of department administrators in reducing manpower on certain types of companies, during "off-peak" hours, or in outlying engine companies. The New York Court of Appeals has upheld the constitutionality of a law requiring binding arbitration for safety employees, and an arbitrator in Cheyenne, Wyo. has handed down a binding decision in a bargaining dispute between the City of Cheyenne and the fire fighter's union which included a 12 percent pay increase, exclusion of the fire chief from the bargaining unit, and a deadline date for the city and the union to settle upon the number of men to serve each station.[50]

Another illustration of the erosion of control involves the right of a fire chief to assign his personnel where he believes the needs of the department will be met best.

Oftentimes the necessity for rotating young fire fighters through busy stations to obtain varied experience on different types of companies and to receive ratings during their training periods from a number of officers conflicts with the claims for assignment preference based upon seniority. In a 1974 dispute involving a member of the Saratoga Springs, N.Y., fire department and the fire fighter's union, the arbitrator ruled that training considerations cannot supersede seniority rights to assignment preferences. Moreover, he ruled that the senior fire fighter had a "stronger" claim to be included in the training program and ordered that he be assigned to the company of his choice part-time during his training, and receive due consideration for permanent assignment after the training was completed.[51]

In New York City an action was brought against the fire commissioner to prevent the closing of eight fire stations. After producing evidence that this action was being taken in accordance with a study made by the Rand Institute to conserve resources and to equalize fire suppression services in the city, the court rejected the plaintiff's argument that the safety of life and property would

---

[48] Evansville v. State, 118 Ind. 426, 21 N.E. 267, 4 L.R.A. 93 (1889).
[49] State *ex rel.* Smith v. Moore, 55 Neb. 480, 76 N.W. 175 (1898); 41 L.R.A. 624, overruling State v. Seavy, 22 Neb. 454.
[50] "Employee-Employer Relations Development," *California Fireman,* Sept. 1975, p. 7.
[51] FIRE DEPT. PERS. REP., July 1975, p. 3.

be endangered, as not involving a constitutional right, and found that these station closings were justified.[52]

Examinations

In recent years there has been a trend away from examinations containing the type of question which determined whether an applicant for entrance to the fire department knew how many bales of hay were allowed in fire district one and similar detailed fire code provisions toward examinations having questions which would reflect the applicant's ability to comprehend written instructions, to apply basic mathematics and mechanics to job related problems, as well as to recognize common tools used in various trades, and otherwise indicate a measure of mechanical aptitude. Such tests are more likely to be accepted as job-related than IQ tests, or the type of examinations which were found to be discriminatory by the U.S. Court of Appeals in Massachusetts. The court found that black and Spanish-surnamed candidates typically performed more poorly on paper and pencil tests of multiple choice type. The court ruled (Griggs v. Duke Power) that a requirement for job entry or promotion that applicants secure a high school education and a passing grade on both the Wonderlic Personnel Test and the Bennett Mechanical Aptitude Test violated Title VII of the Civil Rights Act of 1964.

"The facts of this case demonstrate the inadequacy of broad and general testing devices as well as the infirmity of using diplomas or fixed measures of capability. History is filled with examples of men and women who rendered highly effective performance without the conventional badges of accomplishment in terms of certificates, diplomas or degrees. Diplomas and tests are useful servants but Congress has mandated the common-sense proposition that they are not to become masters of reality."[53]

After the U.S. Supreme Court denied a hearing on the appeal of the above case, the Massachusetts Fire Chief's Association proposed a new physical strength test for fire department candidates. The test would consist of climbing one ladder, crossing over another ladder and descending a third; climbing an aerial ladder at a steep incline and descending; hoisting four times his own weight, then lifting a 40-lb. weight 60 feet using a pulley, but without breaking the upward movement of the weight; attaching hoses to three outlets of a hydrant and extending each hose to its 50-foot length in a predetermined time; carrying 150 feet of 2½-inch hose over his shoulder playing it out, and in return picking up the hose and bringing it back to the starting position; climbing through a long narrow tube; and passing a swimming test, if the local chief could show a need for this ability.

A federal court found that though a Public Personnel Association entrance examination used in Columbus, Ohio, served as a valid predictor of a fire fighter trainee's score on academy exams and instructors' observations, it had not been validated as to the much more important element of work be-

---

[52] Towns v. Beame, 386 F.Supp. 470 (S.D.N.Y. 1974).
[53] See note 16, 401 U.S. 424 at p. 434; 91 S. Ct. 849 at 854.

havior. The court also rejected the department's requirement for a high school diploma or the G.E.D., as not being job related and condemned the practice of giving points for an honorable military discharge for the same reason.[54] However, as pointed out in Alioto v. Waco,[55] the law does not preclude the use of testing procedures, nor require that less qualified have preference over the better qualified simply because of minority origins.

An objective standard is required in giving competitive exams, and it was error to grade an applicant as having failed an oral test consisting of technical questions, where he answered such questions correctly: "A test or examination, to be competitive, must employ an objective standard or measure. Where the standard or measure is wholly subjective to the examiners, it differs in effect in no respect from an uncontrolled opinion of the examiners and cannot be termed competitive."[56]

The author helped conduct the second oral examination in Colorado, of candidates for assistant chief of a large municipal fire department, owing to the fact that the court there ruled that the first oral exam was invalid because the same questions were not asked of each candidate; where the same board members are not used throughout the exam, it would undoubtedly be grounds for declaring the test invalid.

Interview tests have, however, been upheld as valid to appraise the candidate's voice for carrying power, distinctness, and freedom from defects of speech.[57]

In Los Angeles County, California, 3,400 applicants, who had met minimum requirements, applied to take an examination needed to fill only thirty-four vacancies. The court ruled that it is not legal to use a lottery to reduce the number of applicants to 500 for testing. The judge said that he might permit the use of a lottery to pick those who will take the second part of the examination—an oral test—but he said that he wanted to be sure that enough candidates pass the written test so that a lottery will be nondiscriminatory and reasonable.[58] In 1971, when this determination was made, it was estimated that it would cost about $31,000 to conduct the written test for 3,400 candidates, and if they all took the oral test, an additional $46,000.

Eligibility Lists

Before an eligibility list is established there is usually a period of several days to file a protest. If a candidate does not challenge the validity of the list within the prescribed period, in the absence of fraud or collusion, he will be barred from subsequently doing so. However, where a Fort Worth, Texas, fire fighter took an examination for driver in 1972 and placed nineteenth on the eligible list, and though he had filed no protest within the five days allowed after the exam was regraded by the Civil Service Commission (deleting ques-

---

[54] Dozier v. Chupka, 395 F. Supp. 836 (S.D.Ohio 1975); reviewed in FIRE DEPT. PERS. REP., Oct. 1975, p. 9.
[55] Western Addition Community Organization v. Alioto, 340 F. Supp. 1351 (ND Cal. 1972).
[56] Fink v. Finegan, 270 N.Y. 356, 1 N.E.2d 462 (1936).
[57] Sloat v. Board of Examiners, 274 N.Y. 367, 9 N.E.2d 12 (1937).
[58] *Los Angeles Times*, Dec. 11, 1971, Part II, p. 1.

tions which had been improperly included from nondesignated sources) causing him to come out in 21st place, he eventually filed a suit to invalidate the new list nearly a year later when the list was about to expire. The appellate court affirmed the lower court's ruling that he had an equitable property right that was worth enforcing and threw out the revised list.[59]

Manner of Selection

Once an examination had been conducted and an eligibility list established, there was a time when it was fairly routine to select the persons with the highest standing on the list for appointment. That this is no longer true is evidenced by *Fire Command!*, in a June, 1973, editorial, which opined:

"Because of its quasi-military command and organizational structure, the fire service for many years has been somewhat shielded from the give-and-take of personnel qualification and promotion that has developed in many other governmental organizations and in private business. For the most part, fire departments could rely on civil service selection process or, if they were volunteer fire departments, they could practically make their own rules of selection. But today, more and more courts are saying 'No!' and the consequent court orders or injunctions are impossible to ignore."[60]

Notwithstanding the impact of EEOC orders, the affirmative action movement, and some court decisions which seem to cast aside the old merit system, it is possible that once racial balance between minority race members and white members has reached more equitable proportions the courts may once again follow the precedents which have long been established in the selection process.

Under some civil service laws, the person standing highest on the list must be selected. In discussing the charter provisions of the City of Tacoma relative to the certification and appointment of persons under civil service, the Washington Supreme Court said:

"These provisions make it plain we think that the charter requires appointments to be made from the list of eligibles according to relative standings, and that it is the duty of the appointive officer to tender the position first to the person who stands highest on the list, and to the others in turn only after the first has failed to appear on notification, or has otherwise indicated that he does not desire the appointment. The only provision that militates against this conclusion is the requirement that thrice the number of candidates be certified than there are vacancies to be filled, but we think that this requirement was intended rather to secure the appearance of a candidate than to give the appointive officer a right of choice. The evident purposes of the framers of the charter were to make free and open the opportunity to enter the public service, and to secure from the persons applying those shown by tests to be best qualified for the service. These purposes are not accomplished if anything is left to the whim or caprice of the appointive power. There is no danger

---

[59] Crain v. Firemen's & Policemen's Civil Service Commission, 495 S.W.2d 20 (Tex. Civ. App. 1973).
[60] "What is the Best Exam?" (editorial), *Fire Command!*, Vol. 40, No. 6, June 1973, p. 15.

from this construction of the charter of forcing upon the service negligent or incompetent employees. This is carefully guarded against in the charter itself. It is provided that appointive officers and employees shall first be subjected to a probationary term, and power is given to the commissioner in charge to remove an officer or employee, after his term becomes permanent, who is guilty of misconduct, or who fails to perform his duties."[61]

In the above case three names had been certified but only the No. 3 man appeared; the court held that it was error to appoint the No. 4 man, for if only one man appears "it is the plain duty of the officer to appoint that one."

Appointment of Fire Fighters

There was a time when the appointment of fire fighters was generally regarded as purely a matter of local concern. But, in view of the impact of federal legislation, especially with reference to the orders of the Equal Employment Opportunities Commission and court decisions imposing quotas for racial minorities, it can be seen that the appointment of fire fighters has become a matter of national concern as part of the attempt to correct a century of racial discrimination in every walk of life. However, not every court has concurred in a solution of the problem that results in reverse discrimination against the majority races. Nor is race the only grounds on which discrimination has been alleged. Being of the male sex was once a qualification for appointment to the fire department, but no longer; any female who can pass a job-related examination can be appointed, and a number of them have already been admitted to the ranks of the fire service, just as they have to the military academies.

Relevant to this subject of discrimination in appointments is a suit being brought by the American Civil Liberties Union.[62] A young married man was denied appointment to a fire department because during a pre-employment polygraph test, he admitted to having participated in a homosexual act when he was an adolescent, though he had never done so since, a fact corroborated by the test. He maintains that his constitutional rights are being violated, and the suit asks for declaratory relief, a permanent injunction in behalf of the class of persons who would be denied jobs, civil damages, attorneys' fees, and court costs.

Aside from the problem of discrimination, there are other considerations when making appointments to the fire department. For example, where a state law of Indiana listed dispatchers and mechanics among the persons who are members of the "fire force," then the court ruled that the civilians appointed to replace the fire fighters must conform to the statutory requirements as to age and physical examinations.

A minimum height requirement imposed by the District of Columbia Fire Department was held an improper entrance requirement, as not being job related.[63]

In a Kentucky case the plaintiff sought to have the court direct the civil

---

[61] Jenkins v. Gronen, 98 Wash. 128, 167 P. 916 (1917).
[62] Gardner v. Banter, U.S. Dist. Ct. N.D.Ill. (Filed Oct. 1974).
[63] Fox v. Washington, 396 F. Supp. 504 (D.C.C. 1975); see also FIRE DEPT. PERS. REP., Oct. 1975, p. 6.

service examiners of the defendant city to certify his name to the mayor and commission as eligible for appointment as a city fire fighter. The defendants refused to certify him, and instead certified another with a lower grade who was appointed. The court held that the plaintiff had a right to compel the defendants to comply with the ordinance and report the results of the examination, for this was a ministerial duty, but he could not compel them to appoint him.

In the large cities the fire force is generally appointed by the fire commissioners, while in the smaller ones the municipal council generally appoints. Where the fire commission had only the power to make temporary appointments, its appointees were not members of the fire department within the meaning of the city pension provisions for firemen. A board of fire commissioners cannot make appointments to the fire department when the terms of the appointees are not to commence until after the new board is in office.

There is nothing objectionable in vesting in the civil service commission power of removal of names from an eligible list of candidates for appointment, where the city charter provides that names might be removed by the commission after the names have been on the list for two years. But a civil service commission cannot delegate its duty to determine the good moral character of candidates for appointment to the board of fire commissioners. Civil service rules governing the fire department have the force of statutes and are subject to the same rules of construction. Hence, where such a rule required that applicants for examinations (including promotional examinations) be given at least reasonable notice, and the relator was given notice at 7:00 P.M. on the one day that an examination would be held at 7:00 P.M. on the next day, he being on duty from 6:00 P.M. on the day of receiving the notice until 5:00 P.M. the next day, the court held that this was not a reasonable notice, and the relator was entitled to another opportunity for examination.

Civil Service and the Probationary Period

Once the appointment has been made, the candidate must undergo a further test of his qualification for the position—that of satisfactorily passing his probationary trial period prior to permanent appointment. Whether this period is for six months or a year depends upon the applicable civil service rules to any given position.

The civil service law as a whole, and the rules of the civil service commission made under its authority, ought, if possible, to be so construed as to make an effectual piece of legislation in harmony with common sense and sound reason.[64] "The fundamental object of civil service, of course, is to fill governmental positions upon a basis of merit and fitness to serve; and in furtherance of that object many statutes have provided for a period of probation of appointees to governmental offices."[65] It was said in Fish v. McGann:[66] "The object of having a period of probation is to determine whether the conduct

---
[64] Younie v. Doyle, 306 Mass. 567, 29 N.E.2d 137 (1940) (NOTE: *306*).
[65] This entire section is quoted from 131 A.L.R. 384, with minor discrepancies.
[66] Fish v. McGann, 205 Ill. 179, 68 N.E. 761 (1903).

and capacity of the person appointed are satisfactory, notwithstanding the fact that he has passed the exam, and his appointment has been certified by the commission to the appointing office."

In People ex rel. Zieger v. Whitehead,[67] the court said: "The very object and purpose of the probational term is to supplement the labors of civil service examiners in passing on the qualifications and eligibility of the applicant for appointment, and to enable the appointing officer to ascertain and correct any mistake of himself or of the civil service commission arising from the inefficiency of a candidate certified as eligible, when he might in fact prove incompetent to discharge the duties of the place to which he might be appointed. In other words, the probationary term may be said to be *one of the tests* required by statute in determining the efficiency of a candidate. So that until a candidate has taken the test of a probationary term, the appointing power has the right to determine whether the candidate is acceptable, and whether or not he will make the appointment a permanent one."

In People ex rel. Van Petten v. Cobb,[68] civil service made a mistake and put a man's name on the eligible list for appointment without a proper physical exam. The court stated that if the commissioners had made a mistake in placing on the eligible list one not qualified, the act in question afforded a remedy where it provided that "there shall be a period of probation before any absolute appointment or employment...."

People v. Chew,[69] held that even if only one name was certified for appointment, where the rule called for three (but only one passed the exam), he could be appointed. And the court indicated that the probation period would provide the needed chance to evaluate the performance of the man appointed.

In Ptacek v. People,[70] the court ruled that the civil service commission could not open up the exam for assistant superintendent of police to lower grades than the one just below (inspectors), in violation of its own rules limiting promotion to next lower grade.

In a later Illinois case, however, the court refused to correct a similar situation where police captains were allowed to take the exam for assistant chief, when the rules limited it to inspectors; and other cases have allowed the discretion to the Civil Service Commissions of lessening the stated time required to be served in the lower rank.[71]

Promotions—"Freezes"—Discrimination

Where a civil service law requires a certificate of promotion before the increased compensation can be paid, a mere advancement in grade, which automatically entitles the fire fighter, on completion of respective periods of service, to an advancement and appointed compensation without further action by the municipal authorities, is not a promotion.[72]

---

[67] People *ex rel.* Zieger v. Whitehead, 94 Misc. 360, 157 N.Y.S. 563 (1916).
[68] People *ex rel.* Van Petten v. Cobb, 13 App. Div. 56, 43 N.Y.S. 120 (1897).
[69] People *ex rel.* Beach v. Chew, 67 Colo. 394, 179 P. 812 (1919).
[70] Ptacek v. People *ex rel.* Deneen, 194 Ill. 125, 62 N.E. 530 (1901).
[71] People *ex rel.* Witherell v. Chicago, 131 Ill. App. 266 (1907).
[72] Lowery v. New York, 166 N.Y.S. 400 (1904).

Where a civil service commission's regulations provide that official awards (for meritorious acts) shall be considered in relation to promotion only when granted previous to the date of an examination (written or oral), the commission could only consider a candidate's record as it stood at such time, and not as it ought to have appeared (with the record of meritorious service) if it had been properly kept by the board of fire commissioners. Hence, a lieutenant was denied a writ of mandate to compel the commission to give him a rerating on an examination for the position of captain, where for over a year the board failed to certify his meritorious act on his efficiency record.[73]

A city charter provision that the civil service commission shall announce in an examination scope circular the next lower rank or ranks from which promotion will be made vests in the commission a wide discretion, and the court should not interfere with such determination unless it clearly appears that the commission has abused its discretion.[74]

Where a city's budget director refused to issue budget certificates necessary to authorize the fire commissioner to appoint eligibles from the captain's list to the rank of battalion chief, the commissioner "temporarily" promoted 17 captains on the promotional list. After serving a considerable period of time in the higher rank, the men, who complained of not receiving the increased pay, were told that they had to serve in that capacity as long as the financial emergency of the city lasted or else be "brought up on charges." The New York Court of Appeals[75] ordered that the department drop this practice, saying: "There may be no specific statute which invalidates this procedure, but it is totally inconsistent with and subversive of the whole theory of competitive civil service."

Where the City of Detroit, because of economic conditions, imposed a "freeze" on fire department promotions, an arbitrator held that though decisions on filling vacancies was a management function, it was a violation of the union's contract to use "acting" sergeants in fire companies for other than sick leaves, vacations, and short term absences, and the vacancies had to be filled by promotion.[76]

Although the Berkeley, Calif., city charter provides that no appointment shall in any manner be affected by race, color, national origin, or ancestry, it was apparently a feeling of "reverse discrimination" in filling a fire captain's position in 1974 that prompted two lieutenants to bring a class action suit in a fire captain's position in 1974. Two lieutenants brought a class action suit in behalf of eighteen whites who were passed over for promotion in favor of blacks.[77] The judge felt that there was a conflict between the affirmative action program and the merit system, and was critical of the city's policy of scoring written exams only on a pass-fail basis. Of the oral exams he said:

"Determination of one's place in the job opportunity scale is fixed by this vague and undefined oral examination, while the specific, articulated written qualifications tests are substantially ignored."

---

[73] Walsh v. Kaplan, 234 N.Y.S. 159, 134 Misc. Rep. 131 (1929).
[74] Shannon v. McKinley, 62 Cal. App. 2d 169, 144 P.2d 433 (DCA 1st Div. 1944).
[75] O'Reilly v. Grumet, 308 N.Y. 351, 126 N.E.2d 275 (1955).
[76] Fire Dept. Pers. Rep., Nov. 1975, p. 11.
[77] Fire Dept. Pers. Rep., Feb. 1975, p. 10.

In referring to the dilemma of trying to right past wrongs resulting from racial discrimination without perpetuating new wrongs, the judge added:

"New invidious discrimination must not be used to heal the wounds of old invidious discrimination. The Congress and the Supreme Court have made it clear that the cure must not rival the disease by legal sanctuary for a new crop of oppressive procedures, to be harvested by future generations."

A similar view was expressed when Los Angeles County lost a case where reverse discrimination was alleged. A superior court decision in November, 1975 ordered the county to rescind the promotions of a black and a Mexican-American in its library system. The minority candidates had placed eighth and ninth on the promotional list following a competitive examination, whereas the seven plaintiffs who had been passed over included six Anglos and one Japanese-American.[78]

When a promotion is made it is often of great importance to a person on an eligibility list. He may have been tenth on the list when it was established, but after nine appointments have been made he becomes number one. And if a new list is to become effective the next day, and a vacancy occurs on the last day of the old list, he could be appointed before it expires. Though he might be disappointed if he doesn't receive the appointment, the man topping the new list will be greatly pleased, and it is purely discretionary with the chief whether or not he appoints one or the other.

If a dozen vacancies should occur on the last day before an old list expires, the chief would probably want to wait until a new list was certified before making appointments, but not always. Where the Fire Commissioner of New York City created thirty-four new lieutenant positions and restored seventy-two positions that had previously been eliminated, and then filled the vacancies from an eligibility list on the day before its expiration, the court upheld his action, saying that the creation of the positions was lawful and, by way of *obiter dicta*, added that it was within the discretion of the appointing authority the number of eligibles who be promoted and when.[79]

Thus far, no court has been called upon to decide the validity of a requirement that all candidates for promotion or merit raises must, in addition to passing the annual medical examination, also pass a semi-annual physical test, such as is now mandatory in the Salt Lake City Fire Department. However, a New Jersey court has ruled that the condition that one must have served at least twelve months immediately preceding the closing date for filing an application for examination for deputy chief in the permanent capacity of battalion chief is not met by serving over a year as battalion chief, where part of the time served was in an acting capacity or by virtue of a temporary appointment.[80] And the Court of Civil Appeals of Texas has ruled that "immediately preceding," as applied to the requirement that an applicant for promotion must have at least two years of continuous service immediately preceding a promotional examination, excludes an applicant whose service was interrupted by a

---

[78] Los Angeles Times, Part II, Nov. 15, 1975, p. 1.
[79] Love v. Bronstein, 43 App. Div. 2d 426, 352 N.Y.S.2d 457 (1974).
[80] Morse, H. Newcomb, "Legal Insight: 'Acting' Service Doesn't Count," *Fire Command!*, Vol. 41, No. 1, Jan. 1974, pp. 28–29.

period of disability retirement during the two years preceding his application. If the legislature had intended any other exception than the one spelled out relating to persons recalled to military duty, it would have so specified.[81]

Rules and Regulations

Rules and regulations of fire departments are usually adopted to act as a guide for the conduct of the members of the fire department. A breach of these rules sometimes carries a penalty of dismissal, and where such is the case, the modern tendency is to provide the accused with the equivalent of due process, i.e., notice and opportunity to be heard before an impartial tribunal.

Filing a Law Suit Restriction

A rule that requires a fire fighter to obtain permission from his superior officers or the board of fire commissioners to file a civil action is unreasonable, arbitrary, and unconstitutional. A Missouri case held that removal of a person based on a violation of this kind of rule and regulation was improper as contravening the constitutional requirement that the courts of justice shall be open to every person, property, or character.[82]

Going through "Channels"

A rule that requires fire fighters to make all complaints through their superior officers is not a valid restriction on their freedom. A California Appellate Court sustained a chief's termination order where a fire fighter had been suspended for wearing nonuniform shoes, and persisted in taking his grievance (loss of a day's pay) to the city administrative officer in direct disobedience to an order from the chief that he must process his complaint through proper channels.[83]

Criticizing Department Policy

A rule that forbids criticism of fire department policy is also invalid. Free speech is a right which is not abandoned upon accepting public employment, and the submission of a letter to a newspaper critical of the department's personnel policies in violation of the city of Berkeley's personnel rules was not a valid ground for a fire fighter's dismissal. The California Court of Appeals affirmed the superior court's ruling that a public employee may speak freely as long as he does not impair the administration of the public service in which he is engaged; in this instance there was no showing that the fire fighter's conduct impaired the public service.[84]

Generally speaking, public employees have the right to publicize the facts

---

[81] Morse, H. Newcomb, "Legal Insight: Eligibility for Promotional Exam," *Fire Command!*, Vol. 38, No. 9, Sept. 1971, pp. 28–29.

[82] State *ex rel*. Kennedy v. Rammers, 340 Mo. 126, 101 S.W.2d 70 (1936).

[83] Morse, H. Newcomb, "Legal Insight: Willful Disobedience," *Fire Command!*, Vol. 39, No. 11, Nov. 1972, p. 36.

[84] Morse, H. Newcomb, "Legal Insight: Protected Right of Free Speech," *Fire Command!*, Vol. 41, No. 2, Feb. 1974, pp. 36–37.

of a labor dispute. This right was affirmed by the Supreme Court of Iowa in 1967 when it held that the suspensions imposed upon three Cedar Rapids fire fighters for having issued a news release regarding the reluctance of the city to negotiate with their union was a denial of their constitutional right of free speech. In the absence of a showing of impairment of public service they should not have been punished.[85]

Engaging in Political Activities

A rule that forbids fire fighters from taking an active part in politics or engaging in controversy concerning candidates and issues is unconstitutional. Where a fire fighter sought to be elected to his city council in violation of such a rule, the New Jersey Superior Court held that public employees may become candidates for public office and that they may speak freely on public issues and controversies. There was no charge made that the fire fighter was conducting his campaign while on duty or that he was neglecting his job to achieve his political objectives.[86]

The circulation of a petition by fire fighters attempting to get an ordinance adopted that would reduce their working hours was held by the Supreme Court of Alabama not to be "taking part in a political campaign" in violation of a rule forbidding such action except when done privately. The rule also had an exception to the effect that it was not to be construed as denying anyone the right to petition his government.[87]

After decisions by the United States Supreme Court in 1973 upholding the Hatch Act and its state counterparts, wherein it was ruled that federal and state governments could continue to prohibit their employees from engaging in political activities, Congress amended the Act in 1974. The amendment provides that state and local officers and employees may take part in political management and in political campaigns, except that they may not be candidates for elective office. As most local agencies have in some way or another received some federal funds, the Hatch Act may have always had some restraining effect upon political activities by local public employees, as well as state and federal employees. As a result of this new amendment, a state or local officer or employee who is employed by a state or local agency whose principal employment is in connection with an activity which is financed in whole or in part by loans or grants made by a federal agency or organization can now participate actively in political management and campaigns, short of being a candidate himself.

Fire fighters should keep in mind, however, that their state may have regulations restricting political activity, as well as state and local agency restrictions which may apply to public employees. Most of these rules are intended to prohibit activities that are clearly improper, such as soliciting political con-

---

[85] Cook, Vernon, "Free Speech Assured for Iowa Fire Fighters by Recent Supreme Court Decision," *International Fire Fighter*, Sept. 1967, p. 10.

[86] Morse, H. Newcomb, "Legal Insight: Candidate for Public Office," *Fire Command!*, Vol. 36, No. 2, Feb. 1969, p. 45.

[87] Morse, H. Newcomb, "Legal Insight: Can Firemen Circulate a Petition for Signatures?," *Fire Command!*, Vol. 39, No. 6, June 1972, p. 32.

tributions from subordinates or while on agency-owned property. These rules do not usually extend to an employee's conduct in his home or private life.[88]

A Michigan court of appeals decision in 1975 struck down a city of Flint regulation forbidding political activity. The court held that when a public employee's political activities are not related to his job responsibilities, he has the same First Amendment rights as any other member of the public.[89] Although the California courts have upheld the right of public employees to run for public office and to engage in political activity, the Supreme Court of Utah upheld the discharge of a Salt Lake City Fire Department employee who had violated a rule against engaging in political activity. A later attempt to argue the unconstitutionality of the rule in the Federal District Court was unsuccessful, for the court declared that, having presented his claim in the state's courts freely and without reservation, he could not later come into a federal court with the same claims.[90]

Grooming—Hair Styles, Beards

The right of fire departments to regulate the length of hair, mustaches, and sideburns, as well as to ban beards and goatees, has been recently contested in a number of states, and usually unsuccessfully. Based upon some decisions briefly mentioned in the following paragraphs, it would seem that the adoption of regulations which bear a reasonable relationship to the safety requirements of the fire service, and not based purely upon the desire for a military appearance or for sake of uniformity, will be sustained in the courts.

For example, the Appellate Division of the New Jersey Superior Court held in 1973 that the record amply supports the conclusion that safety factors warranted such regulations.[91] A similar conclusion was reached by a United States District Court regarding the San Francisco regulations on beards and sideburns on the basis of the need for breathing apparatus to have a tight seal. After explaining that the right to freedom of expression and privacy are not without limit, the judge asserted that the regulation of such freedom becomes necessary in the public interest for the safety of fire fighters. "There is no requirement that the rules and regulations of the fire department shall be fashioned to meet the pleasure or preference of individual members."[92]

Though a New York court held in a case involving a volunteer fire fighter that hair regulations are not valid if based upon an objective of preventing a loss of public confidence,[93] and though a 1972 decision in Pennsylvania held that hair styles could not be regulated merely to obtain uniformity of appearance alone,[94] a 1974 court ruling[95] upheld hair style regulations of the Phila-

---

[88] *California Fireman,* Aug. 1975, p. 9.
[89] *California Fireman,* Oct. 1975, p. 4.
[90] Morse, H. Newcomb, "Legal Insight: Know Your Courts," *Fire Command!,* Vol. 41, No. 5, May 1974, pp. 28-29.
[91] Mansco v. Town of Irvington, 126 N.J. Supr. 148, 313 A.2d 219 (App. Div. 1973).
[92] "Court Trims Firemen's Right to Wear Beards and Sideburns," *Fire Engineering,* Vol. 124, No. 1, Jan. 1971, p. 32 and Casey, Jim, "Off With the Beards!," *Fire Engineering,* Vol. 124, No. 3, March 1971, p. 33.
[93] Morse, H. Newcomb, "Legal Insight: A Ruling on Hair," *Fire Command!,* Vol. 40, No. 1, Jan. 1973, p. 29.
[94] Black v. Rizzo, 379 F. Supp. 837 (E.D. Pa. 1974).
[95] Mancini v. Rizzo, 379 F. Supp. 837 (E.D. Pa. 1974).

delphia Fire Department which allowed trimmed mustaches, Afros, the Bush, etc., but banned beards, goatees, and long sideburns. The regulations also limited hair length to 1½ in. that would not protrude from under the cap band or helmet. The court concluded that long hair was not only a possible interference with the use of breathing apparatus but also a fire hazard, after one of the plaintiffs set his hair on fire with a match in an attempt to show that hair does not burn. A similar conclusion was reached by a New York Appellate Court[96] and by a court in Florida in 1973, when a union-assisted fire fighter challenged the City of Jacksonville's regulations, which were similar to those of Philadelphia. The court said that it found no constitutional justification for judicial interference with what appeared to be a reasonable and a nonarbitrary safety regulation.[97]

Moreover, a California Unemployment Insurance Appeals Board, in overturning a referee's ruling that a man's long hair did not bar him from receiving unemployment benefits, declared: "We conclude . . . that the claimant by his deliberate actions has voluntarily and materially reduced his labor market. Hence he is not available for work and is not eligible for benefits." The State regulations require that a person must be available for work in order to draw jobless benefits.[98]

"Moonlighting"—Outside Employment

The law relating to dual employment of public employees varies with the jurisdiction, and where fire fighters work on a twenty-four hour shift schedule, often permitting them to be off duty more than twenty days each month, the courts may be more inclined to consider any rule that forbids outside employment as an unreasonable restriction. For example, the Supreme Court of Louisiana, in declaring the invalidity of a City of Crowley's ordinance forbidding "moonlighting" as applied to fire fighters, said: "Admittedly, were the firemen working under a different hourly schedule which would result in a lack of available outside working time, this ordinance might well be reasonable in attempting to insure that firemen were alert, healthy and not lacking in sleep or rest due to outside employment."[99] Likewise, in the case of Harrison v. Chicago Civil Service Commission,[100] the Illinois Supreme Court held that:

> A Civil Service Employee's time does not belong entirely to the city, beyond the fact that he completes his scheduled tour of duty, conducts himself with propriety according to social and moral laws and the prescribed Book of Rules, Regulations and Procedures of a given department and that he makes himself available for duty in any eventuality or emergency.

On first reading, this decision appears to sustain fire fighters' rights to hold outside jobs. On second glance, the average responsible department head will

---
[96] Kamerling v. O'Hagan, 512 F.2d 443 (2d Cir. 1975).
[97] R.Y.V. v. City of Jacksonville, 363 F. Supp. 1176 (M.D. Fla. 1973).
[98] *Los Angeles Times,* March 9, 1973.
[99] City of Crowley Firemen v. City of Crowley, 280 So.2d 987 (La. 1973).
[100] Harrison v. Civil Service Commission, 1 Ill. App.2d 137, 115 N.E.2d 521 (1953).

question whether any employee in an emergency service can "make himself available for duty in any eventuality or emergency" while holding down a regular outside job.

On the other hand, there is a Wisconsin Supreme Court decision[101] on this question which sustains a regulation of the Milwaukee Fire Department prohibiting outside or additional employment.

The foregoing interpretations of law indicate that there is at present no firm nationwide rule on the question of dual employment of city employees.

Endorsing Products

A rule that forbids members to use their names and photos in advertisements without approval of the fire department is valid and reasonable.[102]

Use or Possession of Liquor

The courts have found no reluctance to uphold rules and regulations which forbid drinking alcoholic beverages while in uniform or on duty. Rules which bar the possession or use of liquor on fire department premises are considered reasonable and not an unconstitutional deprivation of one's freedom.

Partaking of Meals Together

Where it had been the practice to have part of the members of a fire station prepare their meals in a common mess club, excluding others, it was held that a rule that requires all members in a fire station to participate in an organized mess is not arbitrary or unreasonable.[103]

Membership in Associations

It has been held that "any association which, on any occasion, for any purpose, attempted to control the relations of members of either the police or fire departments toward the city they undertake to serve, is, in the very nature of things, inconsistent with the discipline which such employment imperatively requires, and therefore subversive of the public service and detrimental to the public welfare." This quotation is from the decision in Hutchinson v. Magee[104] which held that a firefighter who refuses to sever his connection with an association which had organized a strike of city fire fighters may be dismissed from the service if he refuses to do so.

Following a discussion of a recent attempt by the IAFF Union Local No. 2340 to seek redress from an order forbidding Elk Grove, Ill., fire captains and lieutenants from joining the union on penalty of dismissal, the editor of the *Fire Department Personnel Reporter* commented on page 10 as follows, "A fire department cannot forbid fire fighters the right to join a labor union.[105]

---

[101] Huhnke v. Wischer, 271 Wis. 66, 72 N.W.2d 915 (1955).
[102] Kane v. Walsh, 295 N.Y. 198, 66 N.E.2d 53 (1946).
[103] Warner v. City of Los Angeles, 42 *Cal. Rptr.* 502, 231 Cal. App.2d 904 (1965).
[104] Hutchinson v. Magee, 279 Pa. 119, at 120, 122 A.234 (1923).
[105] Melton v. City of Atlanta, 234 F. Supp. 315 (N.D. Ga. 1971); Atkins v. City of Charlotte, 296 F. Supp. 1068 (W.D.D.C. 1969).

However, state laws which exempt 'supervisors' from bargaining laws have been held constitutional. Illinois has no such law, and the National Labor Relations Act, which has a similar provision, does not apply to fire departments. And finally, there is no assurance that a court would hold that fire captains and lieutenants are 'supervisors.' Several state public relations boards have included them within the bargaining units served by IAFF locals." As for the Elk Grove case, the court decided that the union could bring the action under the Civil Rights Act in behalf of the fire department members. A decision on the merits of the contention that prohibiting the officers from joining the union would violate their First Amendment rights was later rendered against the union. The court held that supervisor membership in unions creates a conflict of interest and may be curtailed.[106]

Review Activities

1. The early colonial law that imposed a fine of fifteen shillings on "every person who shall suffer his house to be on fire" was probably the earliest American legislation relating to fire safety.
    (a) Explain why you feel this was or was not a good law.
    (b) Do you think that if fees were imposed on fires today, the increasing number of occupancy fires might be reduced? Why or why not?
    (c) What would you include in a written law to help reduce fires?
    (d) Explain whether or not you feel a fine or other type punishment should be imposed on the owners of occupancies that have caught fire.
2. Outline what you feel would be a successful fire department plan for carrying out a lot cleaning program. Include in your outline descriptions of some of the legal problems you might encounter, and explain how such problems might be counteracted.
3. With a group of your classmates, discuss whether they feel a fire fighter is a public officer or an employee. Then interview some residents and some fire fighters from your community to determine: (1) in which of these ways fire fighters are viewed, and (2) how fire fighters view their own role. Compare your findings with those of your classmates, and write a brief summary explaining whether or not the interviews changed your opinion and the opinions of your classmates.
4. Explain why each of the following statements are true or false:
    (a) The fire chief is the head of the fire department, and the department's board of directors comes under the fire chief's jurisdiction.
    (b) Workmen's compensation covers a fire fighter if injured at a fire, even if that fire fighter has been drinking alcoholic beverages.
    (c) Volunteer fire fighters are not entitled to workmen's compensation if injured while traveling home after a fire.
    (d) Because a fire truck has a red light on it, it is entitled to pass through a stop sign without coming to a complete stop.

---

[106] Elk Grove Fire Fighters Local No. 2340 v. Willis, 391 F. Supp. 487 (N.D. Ill. 1975); 400 F. Supp. 1097 (N.D. Ill. 1975).

(e) All newly appointed fire department personnel should have medical examinations before starting work.
5. Does a fire fighter have the right to carry a weapon? Does a fire officer have the right to arrest someone, such as an arson suspect or a person who has pulled a false alarm? Why or why not?
6. As your community's new fire marshal, you must outline your department's safety regulations for construction and firesafety for a newly constructed light manufacturing occupancy. List the recommendations you would give to the owners of the new building in order for the building to be appropriately protected from fire.

Chapter Six
CITY'S LIABILITY FOR ACTS OF FIRE DEPARTMENT

Conditions Precedent to Municipal Liability

The liability of a municipal corporation for the tortious (wrongful) acts of its agents and employees has been described as an extremely baffling concept arising from the peculiar rules governing municipal liability. Municipal immunity in the case of governmental functions is an archaic doctrine deduced from the "King can do no wrong" theory. The doctrine of *respondeat superior* attaches liability to a municipal corporation for negligence of its officials engaged in a "proprietary" function, or sometimes called "corporate" function, while complete immunity was said to exist if the function is "governmental."

In problems involving municipal liability for tort for personal injuries, except as the following was modified by statute, five questions traditionally had to be answered in the affirmative to impose liability:

1. Was the duty violated connected with a proprietary rather than a governmental duty?
2. Was the negligent person a servant of the city?
3. Was the act not *ultra vires?*
4. Was the offending officer acting within the scope of his authority, or if not, was his act subsequently ratified by the municipality?
5. Was the municipality guilty of negligence, and was the plaintiff free from contributory negligence, and if plaintiff is an employee of city is he precluded from recovery by assumption of risk or the fellow servant rule?

Fire Fighting, a Government Function

For more than a century it was a well established principle of law that fire fighting was a governmental undertaking of a municipality with attendant immunity from liability for torts of its fire department. In the absence of statute the fire department was liable neither for acts of omission or commission, nor misfeasance or nonfeasance.

For a long time this rule of nonliability was constantly followed in many decisions; only the Ohio court dissented from the overwhelming majority. Distinguishing all precedent to the contrary, the Ohio court rejected municipal nonliability for fire department torts by holding that the return from a fire was a ministerial function. In 1922 the Ohio court overruled this unprecedented decision and brought Ohio back into line.

While there had been practically a universal rule of nonliability of a municipality for torts in this country with respect to fire department activity, this immunity historically did not extend to every act of its fire department, e.g., the construction and maintenance of a door to a fire station, the repair of an electric fire alarm system or the maintenance of a fire station. And though the authorities were practically unanimous in holding that a city may not be liable for damages in tort actions of the character indicated, the courts based their decisions on varying reasons.

Bases for Municipal Nonliability in Tort

An examination of the initial cases in various jurisdictions which have passed on the question of the liability of a municipality reveals the reasons impelling the conclusions attained other than the axiomatic statement of nonliability. They may well be classified as follows:

1. A municipality is required by the legislature to establish and maintain a fire department.
2. Powers conferred on municipalities are public in their nature, but are the instrumentalities of the state.
3. If a municipality is not bound to furnish a facility, that ends all question of liability for failure to furnish it.
4. Fire fighters, in performing their duties, are public officers and not agents of the city.
5. The municipality receives no compensation for the use of the fire department which is maintained by taxation for the benefit of the city.
6. Considerations of a sound public policy.

In an Alabama case[1] the court said it was sound public policy to relieve the city of the disadvantages and embarrassments of responsibility for those inevitable miscarriages which attend the performance of duties at once so difficult, so urgent, and so important. "To hold otherwise 'might well frighten our municipal corporations from assuming the startling risk involved in the effort to protect themselves from fire.'"

In Klassette v. Liggett Drug Co.,[2] a serious fire broke out in the defendant's store, and the Charlotte (N.C.) Fire Department pumped water into the building from three engines for over an hour. A woman pedestrian fell on the slippery wet sidewalk in front of the building the following morning and was injured. The court held there was no liability for even if there has been negligence, the woman was contributorily negligent as she could have used another walk. In any case, the fire department was not held liable, because fire fighting is a governmental function and there was no breach of duty by the defendant store owner to the plaintiff, since he didn't put the water there.

---
[1] Long v. Birmingham, 161 Ala. 427, 49 S. 881 (1909).
[2] Klassette v. Liggett Drug Co., 227 N.C. 353, 42 S.E.2d 411 (1947).

The immunity has been recognized when the negligent injury occurred while the fire department was not actually engaged in fighting a fire; e.g., during a practice drill and again while testing an apparatus not yet purchased. The cases are numerous, and at this point it might be well to classify them according to the subject matter involved in the actions, postponing for the moment a discussion of the recent tendency in decisions, as well as by statute, to render municipal corporations liable for their torts irrespective of whether they arise in corporate or governmental functions.

Fire Department Liability

For many years it mattered little how inefficient or unattentive a fire fighter was, or how poorly he fought a fire, because he worked for the "King"—the government—and "the King could do no wrong." Then on January 27, 1961 the California Supreme Court wiped out the old law in the decision of Muskopf v. Corning Hospital District.[3] There it declared that ". . . the doctrine of governmental immunity for tort for which its agents are liable has no place in our law." At the same time, a further in-road to the old doctrine was made in the case of Lipman v. Brisbane School District.[4]

The Muskopf and Lipman decisions have resulted in the overturning of a century of case law which had promulgated the doctrine of immunity from tort liability of public entities engaged in governmental functions. Since fire fighting has traditionally been held to be a governmental function, it can readily be seen that such decisions have an important bearing on the fire service.

While these are California cases, the doctrine of sovereign immunity had also been abrogated in other jurisdictions; e.g., Hargrove v. Town of Cocoa Beach in Florida,[5] and Moliter v. Kaneland Community Unit District in Illinois.[6]

The trend in this field, to which the Muskopf case gave further impetus, has led to the abolition of the doctrine of sovereign immunity by court decisions in Michigan,[7] Minnesota,[8] and Wisconsin.[9] The Arizona Court also repudiated the immunity doctrine.[10]

Thus, we find a trend toward repudiating the ancient doctrine that "the King can do no wrong," both by judicial decision and statute. Where this has occurred, fire fighters should not take for granted that no liability will arise if they permit homes to burn while they are marching in a parade miles away across town or if they become so inebriated that they cannot find the location of the fire, and similar situations described in the early cases where, notwithstanding the negligence of the fire fighters, no liability was imposed. It can no longer be asserted that fire fighting is a governmental function in which there is com-

---

[3] Muskopf v. Corning Hospital District, 55 Cal.2d 211, 359 P.2d 457 (1961).
[4] Lipman v. Brisbane School District, 55 Cal.2d 244, 11 C.R. 97 (1961).
[5] Hargrove v. Town of Cocoa Beach, 96 So.2d 130 (Fla. 1957).
[6] Moliter v. Kaneland Community Unit District, 18 Ill.2d, 163 N.E.2d 89 (1959).
[7] Williams v. City of Detroit, 364 Mich., 231, 111 N.W.2d 1 (1961).
[8] Spanel v. Mounds View School Dist., 118 N.W.2d 795 (Minn. 1962).
[9] Holytz v. City of Milwaukee, 17 Wis.2d 26, 115 N.W.2d 795 (Minn. 1962).
[10] Hermondez v. County of Yuma, 91 Ariz. 35, 369 P.2d 571 (1962).

plete immunity from liability. Whatever immunity exists must be based on statutes, not on common law.

Some states, such as New York, have passed laws which declare public entities liable in tort to the same extent as private persons,[11] though judicial modification has granted a measure of tort immunity for certain types of "purely governmental functions.[12] A statute in the State of Washington[13] is broad enough to constitute a similar blanket waiver of immunity."[14]

At the 71st annual meeting of the Western Fire Chiefs Association (1966), Chief Justice Hugh J. Rosellini of the State of Washington, speaking on the subject of fire department liability, and after mentioning that his state had adopted a law in 1961 which permitted tort actions to be brought against it, observed that other states having similar laws included Alaska, Arizona, California, Florida, Illinois, Kentucky, Michigan, Minnesota, New Jersey, Colorado, and Wisconsin. "It is," he said, "the modern trend in the law to impose liability for negligent acts." He also pointed out that, even in those jurisdictions still granting immunity to a city, it does not extend to employees, "but no one ever bothers to sue an employee of the city because they are so miserably paid, you wouldn't get any money anyway." Later, he warned that the city might try to fasten the blame on a fire fighter who was negligent, for example, in not properly securing a ladder to the truck, and thereby attempt to avoid liability on the ground that the city was not negligent, only the fire fighter, and under such circumstances the fire fighter who violated the standard procedure for securing the ladder could be held liable for any injuries caused by the ladder falling off the truck.

In the Lipman case[15] the California Supreme Court said that, "All that was abrogated was the doctrine of governmental immunity, and its corollary distinction between 'governmental' and 'proprietary' activity."

As previously indicated, and as corroborated by Judge Rosellini, the present trend would appear to be toward repudiating the doctrine of municipal immunity and holding a public entity liable for the negligent acts of its employees regardless of the type of activity in which they are engaged; any immunity allowed hereafter will probably be based upon statute rather than the common law. Some recent cases will be presented to illustrate this.

An example indicating the current viewpoint of the courts on fire department responsibility is the case of City of Fairbanks v. Schiable,[16] an action brought against the city for the death of an apartment building tenant from asphyxiation because of negligence with respect to fire fighting. A woman, Druska Schiable, died of asphyxia during a fire in a building where she and her husband had an apartment. As her executor, her husband brought a wrongful death action under an Alaska statute against the landlord and the City of Fairbanks. The Superior Court rendered a judgment against both for $50,000, and the

---

[11] New York Ct. Cl. Act Sect. 8.
[12] Herzog, Liability of the State of New York for "Purely Governmental Functions," 10 *Syracuse L. Rev.* 30 (1958).
[13] *Wash. Rev. Code* Sect. 4.92.090 (Supp. 1961).
[14] "Abolition of Sovereign Immunity in Washington," 36 *Wash. L. Rev.* 312 (1961).
[15] See note 4.
[16] City of Fairbanks v. Schiable, 375 P.2d 201 (1962).

case was appealed to the Supreme Court of Alaska.

The facts of the Schiable case were that the building was a four-story, concrete structure, and had a dumbwaiter used for carrying garbage cans from upper floors to basement. Ventilating ducts interconnected all the kitchens, bathrooms and clothes closets of each apartment. The fire was started by children near a dumbwaiter door on the second floor; the fire spread to the third floor, burning with great intensity the wooden wainscoting up the stairs and along the hallway, and creating a great amount of smoke. The heat melted the soldered ducts and the grease accumulation burned out. The decedent was seen to collapse at her window following a puff of black smoke from that same window.

The plaintiff argued that the explosion in the duct caused the smoke to come in the decedent's apartment and kill her, but the Court said that that was not necessarily the case—only conjecture—for it may have come in her door, and the landlord should not have been held liable by the lower court on the theory that he didn't clean the ducts or have a safe dumbwaiter shaft.

The court reasoned as follows: Generally a landlord is not liable for injuries to a tenant caused by a dangerous condition which existed when the tenant took possession of the leased premises.[17] Exceptions to the rule apply to the portion of the building over which the landlord retains control, e.g., stairs, halls, elevators, porches, dumbwaiters, and other parts of the premises maintained for the benefit of the tenant and in the purpose of the lease. Where an exception exists to the rule of nonliability, there must first exist a dangerous condition involving an unreasonable risk and, therefore, there is a duty on the part of the landlord to correct the condition. Here the only dangerous conditions were the wooden stairs and trim, open stairs, wood doors on dumbwaiter shaft (unsubdivided ducts with no automatic fire dampers), but these hazards were present when the tenant moved in and were apparent at the time; they were not concealed hazards, and she assumed the risk to the same extent as one would assume the risk of a fire spreading by means of a wooden stairway.[18] No ordinance imposed the duty to construct the shaft in a safe manner, and none was imposed at common law. The court observed that:

> All that the decedent had a right to demand was that these premises be maintained in a condition no more hazardous than existed when she began her tenancy.
>
> "The failure to warn a decedent was never established as the proximate cause of her death, and so the question of whether there was negligence in this respect isn't involved in the case.
>
> "In any event, liability cannot be predicated on this inaction, except possibly where it results from wanton neglect or callous indifference (Stewart v. Raleigh County Bank, 121 W. Va. 181), or perhaps even where one can, with little inconvenience, protect another against a serious peril and fails to do so.

In this case the building manager attempted to fight the fire with the stand-

---
[17] Rest. of Torts, Sec. 356 (1934).
[18] McCall v. Cameron, 126 Minn. 1.44, 148 N.W. 108, 109 (1914).

pipe hose while calling out the alarm on the second floor. He went down the hall on his hands and knees hammering on doors to warn tenants, but couldn't get to the third floor for it was too hot because of the burning wainscoting on the stairway between floors. Of the manager's action the court said: "He did everything a reasonably prudent person would have done in these circumstances. We find no breach of duty owed to a decedent."

On the question of municipal liability the court made the following preliminary observation:

Except for the negligent operation of fire equipment on public streets, where liability has been imposed on the theory of nuisance,[19] it appears to be the rule without exception that a fire department maintained by a municipal corporation belongs to the public or to the governmental branch of the municipality, and that the municipality is not liable for injuries to persons or property resulting from negligence connected with the fire department's operation or maintenance.

However, because of the statutory law in Alaska, the court held that the City of Fairbanks, which maintained a fire department, may be held liable for injuries resulting from negligence connected with the fire department's fire fighting activities. The court stated that the law in plain language imposes liability "for an injury to the rights of the plaintiff arising from some act or omission of the city. There is nothing in the statute which suggests that liability in the operation and maintenance of a city-owned fire department is to be excepted, and we are not justified in reading any such exception into the law."

Moreover, the Court held that the fire department's failure to rescue the decedent, and its affirmative action in preventing others from saving her amounted to negligence and this was the proximate cause of her death. Then the court elaborated on the circumstances, as follows:

"With her escape through the hallway cut off by heat and smoke, she was standing by her open window awaiting rescue. A fire fighter had climbed up a 24-foot ladder he had placed against the building in an attempt to reach her, but it was 8 to 10 feet too short. Efforts were then made to place a 12-foot ladder on top of the 24-foot one, but this was found impracticable. The fire department had one 50-foot ladder but it was being used to evacuate tenants on the other side of the building.

"We do not find that the lack of an additional ladder of sufficient length alone would be sufficient evidence upon which to base a finding of negligence; for other circumstances show that the fire department stepped out of its traditional role of simply doing the best it could to fight the fire and save lives. It assumed a particular obligation to save Mrs. Schiable's life, and then failed to fulfill that obligation by not using other available means to effect her rescue.

"While a fireman was backing down the ladder after an unsuccessful attempt to reach decedent's window, he said to her: 'Don't jump, we'll get you out.' Efforts were then made to find a longer ladder but they were all in use elsewhere. In the meantime no attempt was made to rescue decedent with a fire net. No attempt was made to place an available fire truck against the building

---

[19] Maxwell v. City of Miami, 87 Fla. 107, 100 S. 147 (1924).

and place a 24-foot ladder on top of it in order to gain the additional few feet needed to reach the decedent's window. From ladders too short to reach other apartment windows, ropes had been thrown to other persons awaiting rescue. But that wasn't done for Mrs. Schiable. The fireman who reached the top of the 24-foot ladder, and was only 8 to 10 feet from the window, made no attempt to deliver a rope to Mrs. Schiable. In fact, he couldn't remember whether 'there was ropes on the rig or not.'

"The city was guilty not only of failure to use common sense methods to rescue decedent, but also guilty of affirmatively preventing rescue by others. Some spectators had obtained an extension ladder of sufficient length to reach decedent's window. They raised the ladder and started to extend it, and then were ordered by a city fire inspector to get away from the building. When they refused to obey, they were driven off by fire hoses turned on them by orders from the fire inspector. A bystander named Anderson climbed a ladder, put his head and shoulders in the window, grabbed Mrs. Schiable by the foot and attempted to pull her over to the window. He tried to get a fireman to give him a rope and help him get her out of the window. But instead of getting the assistance he asked for, he was ordered to get off the ladder. It was Anderson's belief that Mrs. Schiable was not dead at that time.

"The evidence is sufficient to sustain a trial court's finding of negligence. This is not merely a case where the court in retrospect and using hindsight had determined that a city might have done things differently in its over-all method of fighting a fire. This is a case where the city specifically induced reliance on the skill and authority of its fire department to rescue decedent, and then failed to use due care to carry out its mission, and excluded others from taking action which in all probability would have been successful. It thus placed decedent in a worse position than when it took complete charge of rescuing her, and became responsible for negligently bringing about her death."

While the Alaska case was somewhat out of the ordinary, there was a lawsuit against Los Angeles,[20] decided in 1962 in favor of the city, in which the plaintiff's barn caught fire and some very valuable horses were lost. It so happened that the nearest battalion chief was out on another fire nearby and had it practically mopped up when he noticed the loom-up a few blocks away; he radioed the dispatcher that he was taking a patrol truck with him to investigate and on arrival he called for a full assignment, then ran back to the barn where the horses were out in a chain-link fenced corral; there was no gate and no way for him to help the horses, for the radiated heat from the large pile of burning hay was intense. By the time the patrol rig laid out its 300 feet of hose and long before the other companies arrived, the barn was a total loss and the horses fatally burned. The owner claimed that the city was liable for damages under the Public Liability statute because the hose was too short by 20 feet to reach his barn, and because the chief had no ax with which to chop down the cyclone chain-link fence.

The court held that the city was not liable because operating fire equipment was a governmental function. Later on, in the landmark case of Muskopf v.

---
[20] Thon v. City of Los Angeles, 21 *Cal. Rptr.* 398 (1962).

Corning Hospital[2] (mentioned earlier in this discussion), the California Supreme Court threw out the governmental function theory of immunity as being mistaken and unjust: "The pre-sixteenth century maxim that 'The King can do no wrong' was merely that the King was not privileged to do wrong. . . . How it became a basis for a rule that the State and Federal governments did not have to answer for their torts has been called one of the mysteries of legal evolution." It goes on to say that: "The rule of governmental immunity for tort is an anachronism without rational basis, and has existed only by the force of inertia . . . no one defends total governmental immunity. In fact, it does not exist. It has become riddled with exceptions both legislative and judicial, and the exceptions operate so illogically as to cause serious inequality."

Summary of Fire Department Liability

Fire fighters in states having adopted laws like those found in the new Government Code sections of California are now granted the needed immunities by express statutory law instead of having to rely upon court decisions—precedent—to achieve immunity from liability. What, then, of the city's liability? If it is greater, and this must be admitted, at least it is more equitable than ever before. For example, under the old "hit and miss" court decision method of fixing liability based upon the distinctions between such nebulous legal concepts as "proprietary" and "governmental" functions, or "ministerial" and "discretionary" duties, we had the unfair situation where a woman could obtain a judgment against the city for injuries sustained by tripping over a door sill or stepping in a small crack in the sidewalk, but a man who lost his home and all his possessions by a fire which "got away" from a weed burning crew could not even state a cause of action against the city because lot burning was considered a "governmental" function.

It is apparent from reading the cases throughout the nation that the acts of public employees are going to be judged on the basis of what any private citizen would consider to be reasonable under the circumstances, and that traditional immunities, unless enacted into statute, will be narrowly construed or denied where necessary to achieve an equitable result to the injured party.

The recent cases discussed on the following pages dealing with collisions of fire apparatus illustrate the tendency of the courts to hold drivers more accountable for their actions under the statutory modifications of the old common law rules.

Fire Apparatus

It was long before the era of streamlined, siren-screaming fire engines when American cities began to defend civil actions for injuries sustained by the negligent operation of fire apparatus. Men were even knocked down by hand-drawn hose carts, and with the advent of the steam fire engine with its all-too-frequent tendency to scare horses and cause them to run away, or to run into buggies and pedestrians, the courts began to advance their various reasons why a municipality should not be held liable. Hence there developed at an early day the municipal immunity for a negligent use or a failure to use its fire apparatus.

At common law, and in the absence of statute, a municipal corporation was not ordinarily liable for negligence of members of a fire department in driving fire apparatus. Until recently the prevailing law throughout the nation was that a municpality is not liable for injury resulting from the negligent mismanagement of fire apparatus, without regard to whether it is being used in the extinguishment of fire or otherwise.

It is well settled that municipalities, in the absence of statute, are not liable for injuries caused by the defective condition of its fire trucks, although a distinction has been attempted in some states between a city's duty to the public and its duty to its own employees in this respect. In a Wisconsin case an attempt to impose liability on the theory that the city was negligent in retaining incompetent fire equipment drivers was unsuccessful.

In states having public liability acts similar to California's there may be liability for dangerous or defective condition of publicly owned motor vehicles.

Various attempts have been made to impose liability on the city for injuries resulting from collisions involving fire trucks on the theory that the operation of a vehicle in a fast or reckless manner constituted an actionable nuisance; and while this contention has been upheld in Florida, it has been rejected in Tennessee, Missouri, and Connecticut.

Notwithstanding that fire engines are performing a governmental function, the city may be held liable where the collision results from a defect in the street or a hole in the street coupled with the fire fighter's negligent operation of the vehicle, but not where the defective condition of the street consisted in a failure to enforce the parking ordinance together with a slippery condition caused by the street sprinkling department.

Fire Trucks—Statutory Modifications of Common Law Rules

Thus, the harshness of the common law rule of municipal immunity has led to some attempt at judicial modification, as in the case of Fowler v. City of Cleveland,[21] and the methods just discussed.

At an early period, many states, including California, held that the general restrictions as to speed and turning corners, applicable to ordinary vehicular traffic, do not apply to fire apparatus responding to an alarm, in view of the fundamental rule of construction that a statute is not applicable to the government or its agencies unless expressly included by name. In some cases it has been held that either an implied exception must be read into local ordinances regulating speed in favor of the fire department or that the ordinance must be declared to that extent unreasonable and void.

Recently many states, like California, have adopted statutes imposing liability on cities for injuries arising out of activities ordinarily considered to be "governmental." But in a Pennsylvania case the statute was held not to alter the common law rule of nonliability as to fire trucks, since a fire truck was not a "vehicle."[22] However, other courts applying more liberal statutes

---
[21] 100 Ohio St. 158, 126 N.E. 72 9 A.L.R. 131 (1919).
[22] Dever v. Scranton, 308 Pa. 13, 161 A. 540, 85 A.L.R. 692 (1932).

have reached different results. Thus in a New York case, the city was held liable for the negligence of a fire fighter operating a municipally owned fire truck in responding to an alarm, even though he was performing a governmental duty.[23] In Miller v. New York,[24] a municipality was held liable under section 282 (g) of the Highway Law for a death caused by the negligence of a fire fighter in operating an auto owned by the fire department.

Illustrative Cases Involving Fire Truck Operation

Cases illustrating the imposition of liability by statute include:

CALIFORNIA—Torres v. Los Angeles:[25] In this case involving the Los Angeles Fire Department, Engine 22 and Engine 14 collided at Jefferson and San Pedro streets on June 1, 1958 while going to a box at 32nd and Trinity streets. All vehicles had their red lights and sirens on and both drivers were familiar with the prescribed routes. Engine 14 was going west on Jefferson and Engine 22 was going north on San Pedro. Each knew the other would be coming. Both engines were being operated by members holding the rank of Auto Fireman, and the driver of Engine 22 had seen Engine 14 pass in front of him many times before (at least 6) on previous runs. The LAFD Rules require that all rigs slow down at intersections, and if a "Stop" light is showing, to stop if necessary, but in any case extreme caution must be used. The stop light was against Engine 22; its Acting Captain was the first to notice 14's engine coming from the right, and its driver, when asked if he ever looked to his right in the last 100 feet of travel, said: "I can't say that I did or didn't." (He had said on his previous deposition, "If the rig had been there, I would have seen it if I had looked.") He also testified that he had not paid any attention to the signal and that "We carry our own signal with us."

Engine 14 hit the right rear of Engine 22, causing it to collide with the Torres auto which had pulled over and parked in obedience to the siren; it in turn struck the Baskett auto, also stopped for the same reason. The collision resulted in the death of Mrs. Torres, and injuries to Mr. Torres, their daughter, Edith, and both Mr. and Mrs. Baskett. The jury rendered judgments for plaintiffs.

The city attorney argued on the basis of the old Lucas case[26] that sounding the siren was enough, but the court did not agree. The current state vehicle code, which was in effect in 1958, specifically provides certain exemptions and, unlike the 1938 edition, is so worded that all vehicles, including emergency vehicles, must comply with its provisions subject to those exemptions granted authorized emergency vehicles in the code.

The court adopted the view expressed in Peerless Laundry Services, Ltd. v. City of Los Angeles[27] that the city could be held liable on the general principles of negligence. The vehicular regulations from which exemption is granted related to speed, right of way, pedestrian duties, street car and safety zones,

---
[23] Snyder v. Binghamton, 245 N.Y.S. 497, 138 Misc. Rep. 259 (1930).
[24] 235 App. Div. 259, 257 N.Y.S. 33 (1932).
[25] 58 Cal.2d 35, 372 P.2d 906 (1962).
[26] Lucas v. Los Angeles, 10 Cal. App.2d 476, 75 P.2d 599 (1938).
[27] 109 Cal. App.2d 703, 241 P.2d 269 (1952).

stopping, standing and parking. There is no blanket exemption of municipalities from all the rules that govern the operation of motor vehicles.

The court stated its belief that the correct rule is expressed by the Supreme Court of Wisconsin in Montalto v. Fond du Lac County[28] which held that drivers of emergency apparatus, while not judged by exactly the same yardstick as the operators of conventional vehicles, are bound to exercise reasonable precautions against the extraordinary dangers of the situation that the proper performance of their duties compel them to create. The terms "negligence" and "reasonable care" are *relative* terms and their application depends upon the *situation* of the parties and the degree of care and vigilance which *circumstances* reasonably impose. It cited the case of Horsham Fire Co. v. Fort Washington Fire Co.[29] in which the court said: "The object of a fire truck's journey is not merely to make a show of rushing to a fire, but actually to get there. If the driver is to ignore all elements of safe driving at breakneck speed through obviously imperilling hazards, he may not only kill others en route, but he may frustrate the whole object of his mission and not get there at all!"

The Supreme Court of Nevada in Johnson v. Brown[30] was quoted with approval as follows: "It is clear to us that the majority and better rule in opposition to the (old) California rule is expressed in the Lucas case, *supra*, which requires the driver of an emergency vehicle answering an emergency call to exercise reasonable precautions against the extraordinary dangers of the situation which duty compels him to create, and he must keep in mind the speed at which his vehicle is traveling and the probable consequences of his disregard of traffic signals and other rules of the road." The Nevada court continued: "While we are not unmindful that the Reno ordinances and statutes are designed to give emergency vehicles extraordinary rights, they were not intended to absolve the drivers thereof from the duty to be on the lookout at all times for the safety of the public whose peril is increased by their exemption from the rules of the road. We believe that sound public policy requires such a construction." To the same or similar effect, see Russell v. Nadeau;[31] City of Miami v. Thigpen;[32] Henderson v. Watson;[33] Ruth v. Rhodes;[34] and Grammier-Dismukes Co. v. Payton.[35]

The court also cited a Michigan case on the meaning of "drive with due regard for the safety of others," which also rejected the old California view expressed in the Lucas case: "Driving a fire truck into an intersection in full reliance upon the right to exceed speed limits and the right to proceed instead of stopping for the stop sign or the through street, but without observing or giving any heed to oncoming traffic on the intersecting through street did not amount to driving with due regard for the safety of others as required by statute."

---

[28] 272 Wis. 552, 76 N.W.2d 279 (1956).
[29] Horsham Fire Company No. 1 v. Fort Washington Fire Company No. 1, 383 Pa. 404, 119 A.2d 71 (1956).
[30] 75 Nev. 437, 345 P.2d 754 (1959).
[31] 139 Me. 286, 29 A.2d 916 (1943).
[32] 151 Fla. 800, 11 S.2d 300 (1942).
[33] 252 S.W.2d 811 (Ky. 1953).
[34] 66 Ariz. 129, 185 P.2d 300 (1942).
[35] 22 S.W.2d 544 (Tex. Civ. App. 1929).

The California Supreme Court (at p. 51 in the Torres case) said: "The question to be asked is what would a reasonable, prudent, emergency driver do under all the circumstances, including that of the emergency."

Other California cases include Cavagnaro v. Napa (inadequate brakes and negligent operation of equipment, under Code);[36] Issacs v. S. F.;[37] Johnson v. Fontana Co. Fire Protection District (fire protection district liable under statute);[38] and Dillenbeck v. Los Angeles (training bulletins relating to entry of an intersection against the light, admissible in collision suit).[39]

COLORADO—Delong v. City and County of Denver:[40] In this case a police vehicle responding to an emergency went through a red light at 30 miles per hour and collided with the plaintiff. The court held that departmental rules were admissible which forbid going through a red light at more than 15 miles per hour as standard of care for the driver's conduct, and the court added that an "... employee's failure to follow the safety rules (of his employer) constitutes evidence of his negligence." The case was being appealed to the Colorado Supreme Court and has been remanded for a new trial.

FLORIDA—Thigpen v. Miami:[41] Under wrongful death act, for injury to motorist's child passenger.

ILLINOIS—Bryan v. Chicago:[42] Contributory negligence as defense.

KANSAS—Kansas City v. McDonald:[43] Insurance policy of fire fighters as not affecting city liability.

MICHIGAN—Holser v. Midland:[44] Statute abolishing governmental function as defense.

NEBRASKA—Opocensky v. Omaha:[45] Modifying rules set out in Gillespie v. Lang, 35 Neb. 34.

NEW YORK—Fink v. New York:[46] In this case the fire truck was traveling on the wrong side of the street, sounding its siren and bell, when it struck a deaf mute who was crossing the street with the green light in his favor. A witness testified that the fire department was proceeding at an excessive speed right up to point of contact with the decedent. The court held that the fire apparatus operator was negligent and that the decedent was not contributorily negligent.

OHIO—In a case where the fire chief of the Village of Arcadia collided with another vehicle at an intersection and was charged with not yielding the right-of-way to the other vehicle in obedience to a stop sign, the court stated that the chief's claim that he had the right of way by virtue of using his red light and siren was an affirmative defense. The fire chief satisfied the court that he had complied with the four requirements of the Ohio Vehicle Code which gave him

---

[36] 86 Cal. App.2d 517, 195 P.2d 25 (3rd Dist. 1948).
[37] 73 Cal. App.2d 621, 167 P.2d 221 (1946).
[38] 15 Cal. App.2d 380, 101 P.2d 1092 (1940).
[39] 69 Cal.2d 472, 72 *Cal. Rptr.* 321, 446 P.2d 129 (1968).
[40] 34 Colo. App. 330, 530 A.2d 1308 (1974), aff'd—— Colo.——, 545 P.2d 154 (1976).
[41] See note 32.
[42] 371 Ill. 64, 20 N.E.2d 37 (1939).
[43] 60 Ka. 481, 45 L.R.A. 429, 6 Am. Neg. Rep. 67 (1899).
[44] 330 Mich. 581, 48 N.W.2d 208 (1951).
[45] 101 Nebr. 366, 163 N.W. 325 (1917).
[46] Fink v. New York, 206 Misc. 79, 132 N.Y.S.2d 172 (1954).

the privilege of disregarding the stop sign and dismissed the case against him.[47]

PENNSYLVANIA—Omeh v. Pittsburgh:[48] In this case a fire truck was "responding" to a dump fire when it collided with defendant's pick-up truck. The defendant recovered damages for injuries to him and his family and for damage to his truck. The court held that whether or not the fire fighters were "responding" to an alarm, as contemplated by the law (which is very similar to the California law), was a question of fact for the jury.

The facts outlined in the report of the case indicate that the fire fighters were responding to a "still alarm" in the form of a telephone request to go to a dump fire that had been burning for days; they knew that previous shifts had been wetting down this dump fire, that no structures were exposed, and that there was no need for other than ordinary speed in merely going to relieve another crew.

The fact that the captain, without calling for another company for assistance, ordered his crew to lay a line for a "wash down" and ordered them to extricate the children out of the truck, shows that he knew there was no urgency to get to the dump.

Therefore the court held that the instruction to the jury was correct in that the fire truck must have been responding to an emergency call in order for the city to claim the right to an exemption from liability for anything other than recklessness. Under the conditions present, the city would be liable for ordinary negligence in the operation of the fire vehicle. "The exemptions of a fire department vehicle, under the Vehicle Code, are conditional upon the vehicle being operated with due regard for the safety of all persons using the highway; they do not protect an operator from the consequences of a reckless disregard of the safety of others or from the consequences of an arbitrary exercise of the right of way."

While the question of whether a fire department vehicle is traveling "in response to an alarm" is one of law as to the applicability of the exemption provision to the credible factual proofs where the evidence in such connection, or any material portion of it, is oral, the question is necessarily one for a jury's determination.

"Where the operator of a fire department vehicle knows that there is no necessity to reach the scene of a fire at high speed, or to operate the truck with other than ordinary care, the plaintiff is not required to prove reckless operation of a truck in order to affix liability upon the municipality; in such a case proof of negligence is sufficient."

Statutory Liability for Negligent Operation of Emergency Vehicles—California Example

In considering statutory liability for negligent operation of emergency vehicles the California law is used to illustrate the problems created by this kind of legislation. The laws of other states naturally vary somewhat from those of

---

[47] Morse, H. Newcomb, "Legal Insight: Chief Drives Through Stop Sign," *Fire Command!*, Vol. 41, No. 9, Sept. 1974, pp. 32, 33.
[48] 387 Pa. 128, 126 A.2d 425 (1956).

California, but many of the same problems exist in all states and the legislation is not so different but that California's experience is of real value.

The California legislature has imposed liability on the state and every city or other public corporation owning any motor vehicle for damage by reason of death or injury to persons or property resulting from the negligent operation of such motor vehicles by any officer, agent or employee, "or as the result of the negligent operation of any other motor vehicle by any officer, agent or employee when acting within the scope of his office, agency or employment...."[49]

The Vehicle Code also provides for the subrogation of the city, or other public corporation against which recovery is had, to all the rights of the injured person against the negligent officer, agent, or employee.

The courts have held that in order to recover against the city under the Vehicle Code, it must be shown that at the time of the accident the driver of the city vehicle was acting within the scope of his authority. On a "governmental function" theory, the court held that there was no municipal liability to a magician who was injured when a Los Angeles police captain, off duty, took his wife out to dinner in a police car, and negligently drove into the magician's auto, for the policeman was acting outside the scope of his employment.

But in a case where a Los Angeles Department of Water and Power employee was using a city-owned car on his lunch hour when the accident occurred, the city was held liable for the driver's negligence because the car was owned by the city in its proprietary capacity; the finding that, while at lunch, the employee was not acting within the scope of his employment was no bar to liability under the imputed negligence statute.

In another case the court found that the driving of a Los Angeles fire commissioner from his home to his office, under his captain's orders, was within the scope of the fire fighter's duty, and hence municipal liability was properly imposed. The city had contended that the transportation of a city official to and from his home was unauthorized and unlawful. The court did not deny this allegation, but found that the commissioner had gone immediately from his office to inspect a fire station, and that this had the effect of giving authorization to the whole trip.

Before liability can be imposed under the California statute, the injury must have been caused by the negligent operation of the vehicle, and not, for example, by a dangerous condition of the highway, such as might be caused by an oil spot.

The statute has been held to apply to governmental functions, and also to impose liability on a city as applied to police cars when not on an emergency call. In order to properly determine the question of liability for negligent handling of fire trucks, cases involving all types of emergency vehicles must be considered.

Since emergency vehicles were not specified in the California law imposing liability for negligent operation of vehicles, and since members of police and fire departments are exempt by statute from civil liability when driving author-

---

[49] Cal. State, 1929, p. 565, Civ. Code. Sec. 1714½ as amended by Cal. State, 1931, p. 168; superseded by Cal. State. 1935, p. 93, Cal. Veh. Code. Sec. 400.

ized emergency vehicles, thus preventing any right of municipal subrogation against negligent operators of such vehicles, it was held in an early case that the statute was not intended to include liability for the negligent operation of authorized emergency vehicles.

Later, this view was rejected in a decision in which the court declared, after holding that a chartered city was liable to a person injured by the negligent operation of a police ambulance on an emergency call, that: ". . . the legislature recognized that the members of fire and police departments in the performance of duties are often called upon to exceed the ordinary speed limit and take the right of way. If such employees were compelled to weigh the chances and assume a personal responsibility, speed and efficiency would be affected.

"The legislature, however, realized that such operation of motor vehicles on the streets did create a hazard to the general public and provided that the city should be liable when the exercise of governmental functions resulted in injury to the people. The legislature, also at the same time, authorized its governmental agencies to obtain insurance to indemnify itself against loss by such actions.

"If, too, an employee is unduly reckless or careless in the operation of a vehicle under his control, the municipality still retains its rule making and disciplinary power and can . . . exercise restraint over its employees. It cannot be argued that because the state deemed it expedient to waive its right of subrogation against its employees, it thereby denied a right of action against itself."

The California Supreme Court in 1938 declared the above to be the correct rule in a case where the plaintiff sued for personal injuries received when a fire chief, responding to a fire alarm, drove a city car at a high speed through a boulevard "stop" and collided with the plaintiff's car in the intersection. It declared that under the statute a city is liable to persons injured by the negligent operation of its authorized emergency vehicles. It added that the legislature excused the drivers of emergency police and fire apparatus from personal liability in the interest of speed and efficiency, and, because of the contingent danger to the general public, makes the city liable when the exercise of its governmental function results in injury to the people.

There is a difference between the standard of care applicable to drivers of authorized emergency vehicles and to motorists generally. When responding to an emergency call with a red light to the front and when sounding the siren as a warning, such vehicles are exempted from the general rules of the road and are given the right of way. But the law provides that the exemptions shall not relieve the driver of an emergency vehicle from the duty to drive with due regard for the safety of all persons using the highway, nor protect any such driver from the consequences of an arbitrary exercise of the privileges. Hence, every California fire fighter, and fire fighters located in states having similar statutes to those of California, should note carefully the five conditions which must exist in order to render the city immune from liability for damages inflicted by fire department vehicles:

1. The driver must be operating an "authorized emergency vehicle."
2. He must be responding to an emergency or an alarm of fire.
3. The siren must be sounded, as reasonably necessary.

4. Red light must be displayed to the front.

5. He must drive with due regard to the safety of all persons using the highway, and must not arbitrarily exercise his privileges.

The above points will be dealt with in the following paragraphs:

1. **The driver must be operating an "authorized emergency vehicle.** The mere fact that he is a fire fighter or a policeman does not make his own automobile an emergency vehicle. Such a vehicle should be equipped with a siren and red lights to the front. Thus, under the California Vehicle Code vehicles used by members of a fire prevention bureau or by fire fighters detailed to bureaus, such as hydrants, photography, public relations, etc., could not be classified as authorized emergency vehicles, even though owned publicly and equipped with sirens and red lights, where they are not used in responding to emergency calls. On the other hand, the private car of the chief or the assistant chief of an organized fire department, equipped with siren and red light and used in responding to emergency fire alarms, is an authorized emergency vehicle.

2. **Vehicle must be responding to an "emergency call" or alarm of fire.** One case indicates that it need not be an "emergency in fact" if the driver is responding to what he reasonably believes is a genuine emergency or fire alarm. An ambulance was merely carrying a man to the hospital for an appendicitis operation when it collided with the co-defendant and injured the plaintiff. The court held that the fact the operation was not performed for hours after the arrival at the hospital was inadmissible as immaterial to the question of whether the ambulance was on an emergency call. It reasoned that the test for determining whether a publicly owned ambulance was an "emergency" vehicle within the statutory exemption from the rules of the road is not whether an "emergency in fact" existed, but whether it was being used to respond to an emergency call at the time. Hence it should be immaterial to the issue of municipal liability whether the call should subsequently prove to be a false alarm or not.

Testing a fire car is not such an emergency nor is merely responding to a call to "come to my office."

Does an emergency call include rescuing a person from a cave-in, removing a cat from a telephone pole, resuscitating a drowning victim or a new-born baby, pumping water out of a cellar, or covering a roof with a salvage cover? All these and many other kinds of calls are not uncommon to the modern municipal fire department. One fire department was criticized for failing to respond quickly enough to pull a blanket out of an obstructed sewer. (But without resultant liability to the city for the water damage ensuing.)

If, to be classed as an emergency vehicle, the apparatus must be responding to a call which is within the fire department's authority to answer, and if the authority given by charter extends only to extinguishing and preventing dangerous fires, as in the case of many cities, an interesting query presents itself as to whether a publicly owned fire department vehicle, such as a rescue company apparatus, is an "authorized emergency vehicle" when it responds to an emergency call to save the life of a new-born child. This phase of the fire department's activity might be considered unauthorized in so far as its activities extend to matters entirely unrelated to preventing or extinguishing dangerous fires. However, in the author's opinion no doubt should exist that any rescue company activity at the scene of the fire, whether for the benefit of fire fighters or civilians, in the nature of first aid or resuscitation, is clearly within the implied powers belonging to the fire department, as an incident to its granted powers.

If a court should hold that a rescue apparatus when responding to a respiratory case at a private home, or a municipally owned fire department salvage apparatus when responding to a call to place a rubber cover on a leaky roof, were unauthorized emergency vehicles, because they were not acting within the scope of the authority extended to the fire department by the city charter, the drivers would not be entitled to the exemptions from the rules of the road, nor would they be entitled to the immunity from civil liability afforded to them by state law. Under such a holding the victim of a collision caused by the operation of such a vehicle in disregard of traffic regulations could recover damages from the driver. It is problematical whether the city would be held liable if the city could show that one of the conditions precedent to liability in the case of fire department employees is that they act within the scope of their duty. It could be argued, however, that any activity on the part of the fire department other than the charter-authorized functions of preventing and extinguishing fires is nongovernmental, and hence renders the city liable in its proprietary capacity, just as some courts hold that the maintenance of fire stations or alarm systems by the fire department is a corporate (proprietary) function.

3. **The siren must be sounded.** Some confusion has arisen over the use of an expression in vehicle codes requiring the operator of an authorized emergency vehicle to "sound an audible signal by siren," or, "upon the immediate approach of an authorized emergency vehicle giving audible signal by siren." It is the sounding or giving of the audible signal that fixes the right of way. Where there is dispute as to whether the party injured heard the signal it is a question of fact to be determined by the jury, but only insofar as it is material to the issue of contributory negligence. The statutes are clear in stating that when an audible signal is given, the operator of the emergency vehicle has a clear right-of-way. The giving of the signal is the measure of care on his part, and if this is done his duty of care is perfected, subject to the limitation of "arbitrary conduct," said the court in a Los Angeles case. Here the operator of the car in which the plaintiff was a guest saw the police car approaching the intersection, sounding a siren, and though he could have stopped in time to avoid the collision he thought "he could beat it across the intersection." In holding that inasmuch as a warning siren had been sounded, negligence of the police driver could not be predicated upon his failure to obey the stop signal or his fast rate of speed. The court said: "Our conclusions from the foregoing are that when the operator of an emergency vehicle responding to an emergency call gives the statutory notice of his approach the employer is not liable for injuries to another unless the operator has made an arbitrary exercise of these privileges."[50]

---

[50] See note 26.

In Stone v. San Francisco,[51] a failure to sound a siren by a policeman on the way to a fire constituted an arbitrary exercise of driver's privileges. Nor was a mere blowing of the horn on a vehicle not equipped with a siren sufficient to relieve the City of Beverly Hills of liability for an arbitrary exercise of the privileges of exemption from rules of the road by a policeman.[52]

In those states, such as California, where the vehicle code requires that the driver of an emergency vehicle must sound the siren "as necessary" during emergency response, it is sometimes the practice to withhold sounding it when proceeding on a multi-lane highway to obviate the necessity for the drivers in all the lanes of traffic from having to pull over and stop upon hearing a siren, as required by law. In a recent case where a county sheriff's vehicle, responding on the freeway without use of warning devices, was involved in an accident, one cause of which was the negligence of the other vehicle's driver in changing lanes, the court held that the exemptions from the rules of the road are not available where no red light and no siren are used.

According to the Attorney General of California: "In the absence of specific legislation authorizing the use of an air horn in conjunction with a siren, its use might subject the user to liability for damages should a jury, exercising its right to determine factual issues, find that such use of the air horn removed the right to rely upon the exemptions accorded to operators of emergency vehicles." Electronic "yelpers" are not sirens, and are subject to the same objection as air horns.

4. **A red light must be displayed to the front.** If the red lights are turned on by the operator, the fact that they are rendered invisible by murky weather or are not seen because of the other driver's inattention should be immaterial. The light must be continuous to the front. A single "blinking" or revolving light would not meet this requirement.

5. **The operator must drive with due regard to the safety of all persons using the highway, and must not arbitrarily exercise his privileges.** "Due regard for the safety of all persons" means that the driver must, by suitable warning, give others a reasonable opportunity to yield the right-of-way.[53] "All persons" includes pedestrians, and the latter are under no obligation to anticipate the presence of fire apparatus on the highway unless they know, or by reason of an audible signal, should know that a fire truck is approaching.[54] The court stated in the Los Angeles case: "The expression 'arbitrary exercise of privileges' has caused some confusion. Since the vehicles are excluded from the restrictions of speed and right of way, negligence cannot be predicated upon those elements if proper warning had been given. These are among the 'privileges' which are granted by the statutes. An arbitrary exercise may rest upon the question of whether an emergency in fact existed. If these privileges are exercised in returning from a fire or for some private purpose of the operator, it might be a case of an arbitrary exercise of them."[55]

---

[51] 27 Cal. App.2d 34, 80 P.2d 175 (1938).
[52] Altman v. Beverly Hills, 40 Cal. App.2d, 105 P.2d 153 (1940).
[53] Raynor v. City of Arcata, 11 Cal.2d 113, 77 P.2d (1938).
[54] Eddy v. City of Los Angeles, 28 Cal. App.2d 89, 82 P.2d 25 (1938).
[55] See note 26.

The view that the practical purpose of the phrase "arbitrary exercise of privileges" was to cover cases in which drivers of emergency vehicles, by virtue of the character of the vehicle and by use of its siren, maintain non-necessary speeds and drive with abandon of the usual rules of the road, when there is no need has been advanced.[56] This is the case where a fire engine is returning from a fire in the absence of any new call or emergency; where the vehicle is being warmed up or tested; or where the chief is being rushed to his dinner. One might add to these situations, one where the operator had given the required signal, but saw that another had neither heard nor heeded it. Under these circumstances, it would no doubt be considered an arbitrary exercise of the privileges to continue on into an inevitable collision. This situation might raise an issue of fact similar to that involved under the "last clear chance" doctrine; the negligence of the operator of the emergency vehicle would then rest solely upon his arbitrary exercise of the privileges and not upon his disregard of the ordinary speed and right-of-way restrictions.

In a San Francisco case the "arbitrary exercise . . ." was held to be similar to "wilful misconduct," as in the ordinary guest cases (cases which hold that "guests" riding in automobiles cannot recover from their host driver, ordinarily, unless the driver is intoxicated or guilty of wilful misconduct [more than merely negligent]), which, in another case was held to mean "something different from and more than negligence, however gross."[57]

Most vehicle codes do not, however, make an "arbitrary exercise of privileges" a condition precedent to municipal liability for negligent operation of police and fire vehicles. Actually, it is only one of the previously discussed five grounds upon which negligence in such circumstances can be predicated. So despite the court's definition of the phrase, there will still exist municipal liability, even in the absence of any "wilful misconduct."

The typical vehicle code provision merely sets a different standard of care for the drivers of emergency vehicles, making it not unlawful to exceed speed limits, and generally to preempt the right of way. The vehicle code's admonition for emergency vehicle drivers to show due regard for the safety of others on the highway and not to arbitrarily exercise privileges are not conditions precedent to the exercise of the exemptions, but they are conditions subsequent upon which the drivers must take the "consequences."

The fact that a fire fighter is entitled to exercise the statutory exemptions does not bar municipal liability for his negligent operation of the vehicle. Hence, his failure to slow down and look both ways to observe whether the public is heeding his warning signal would constitute a negligent operation within the meaning of the vehicle code, even though such conduct would not fall within the definition of "wilful misconduct"; vis., "the intentional doing of something either with the knowledge that it is likely to result in serious injury, or with a wanton and reckless disregard of its consequences."

It is a misconception to say that the city would not be liable for "ordinary negligence," but only for "wilful misconduct" in the case of emergency ve-

---

[56] David, L., *Municipal Liability for Tortious Acts and Omissions,* 1936, p. 141.
[57] Stone v. San Francisco, 27 Cal. App.2d 34, 80 P.2d 175 (1938).

hicles. The city is liable for "ordinary negligence," but in the case of emergency vehicles the standard of care is different from that of nonemergency vehicles. The city is still liable for negligence, but the test becomes, "whether the fireman driver exercised the care that a reasonably prudent driver of like emergency vehicles would have exercised under like circumstances," and recognition should be given to the fact that the mere circumstance of not stopping at a boulevard "stop," or traveling at a fast speed, for instance, is not of itself negligence because of the exemptions accorded to emergency vehicles. It would only become so when one or more of the five conditions rendering a municipality immune from liability for damage was lacking.

The question arises of whether or not failure on the part of a fire fighter to perform any one of the five conditions for municipal immunity that results in liability on the part of the muncipality would give the municipality the right of subrogation against the fire fighter. It is logical to suppose, in view of the fact that the immunity provided in the Vehicle Code extends only to fire fighters and policemen, that the right of subrogation given the city extends only to city employees or drivers of city owned vehicles, other than policemen and fire fighters. On the other hand, what are the consequences from which the fire fighters are not protected? Perhaps, in view of the utterance of the court to the effect that fire fighters and policemen are not personally liable in such situations (the above question not having been raised), the imposition of disciplinary measures with the possibility of demotion, suspension or removal might well be regarded as the "consequences" to which involved fire fighters would be subjected.

From this discussion and review of the cases interpreting the California statute regulating municipal liability with respect to emergency vehicles, it can readily be seen that such statutes are a fertile field for controversy. Even the courts have deplored the unjust results that may be reached under these statutes, and they are not universally looked upon with favor. However, the validity of such statutes has been sustained, and it must be remembered that such laws do not render the fire fighters personally immune from liability in the absence of specific provisions to that effect.

Fireboats

Many fire departments today have fireboats in service, and so it is important to note that the old common law rule of nonliability for fire department functions does not extend to this branch of the service. In courts of admiralty, under the maritime law, the city is liable for the negligence of its fire fighters in operating its fireboats. At an early date the United States Supreme Court, in the case of Workman v. New York,[58] where the "New Yorker," a steam fireboat, collided with the barkentine "Linda Park" while the former was running into a slip for the purpose of extinguishing a fire in a warehouse, declared that "under the general maritime law, where the relation of master and servant exists, an owner of a vessel committing a maritime tort is responsible under the

---

[58] 179 U.S. 552, 45 L.Ed. 314 (1899), discussed in HARV. L. REV. 451 (1901), "Municipal Liability for Torts of Fire Boats."

rule of respondeat superior." But in Thompson Navigation Company v. Chicago,[59] where a fire tug of Chicago called the "Yo Semite" negligently collided with the boat "City of Berlin," the Federal Court said: "At admiralty the vessel committing the unlawful injury is the offender, and the owner is liable to the extent of his interest in the vessel; not because he stands in the relation of principal or master to the crew, but alone because of the fact of ownership."

A California case held that a fireboat was not an authorized emergency vehicle and therefore was not exempt from speed limits.

Sending Fire Trucks Outside the City Limits

Many cities make a practice of sending men and apparatus outside the city boundaries in order to assist neighboring city or county fire departments. Although mutual assistance is generally rendered in a spirit of good will and reciprocity (based largely on "gentlemen's agreements"), there is a recent tendency for cities to enter into mutual aid compacts, and in some cases a fee is charged based upon the number of men sent in, the fire trucks used, and the time spent at the fire. These mutual aid contracts are not considered binding unless the officials have the authority to make such contracts granted by the state laws, city charter provisions, or municipal ordinances authorized by state laws or constitutional provisions. The general rule is a municipal corporation can exercise only such powers as have been conferred upon it in its charter or by some general law. The extra territorial jurisdiction of a municipal corporation in California is stated in Ex Parte Blois (1918): "Municipalities may exercise certain extra-territorial powers when the possession and exercise of such powers are essential to the proper conduct of the affairs of the municipality."[60]

There cannot be said to exist in a public official the implied power to bind the city without showing some express authority conferred upon him by statute or constitution. If the officer later receives such authority and then ratifies the contract, or if some board or officer having the power ratifies, or if the subsequent conduct of the authorities constitutes a ratification, then it is binding. Therefore no liability for breach of contract would exist for a failure to respond to a fire call in a neighboring city if the official signing the agreement did so without proper authorization. The implied power on the part of the county to enter into a contract with the city to furnish county fire protection was denied in a Kentucky case, and the contract was held not binding on the city.[61]

However, in Wisconsin the court held that cities have, by their general charters, power to contract and to furnish fire service to people and communities adjacent to the cities, and at least within their trading areas. In this case, the ladder on a truck to which a volunteer fire fighter was holding came loose on the return from a fire from outside the city limits, causing the fire fighter to fall and injure himself. In holding that he was entitled to workmen's compensation from the city since he was acting under a plan entered into by it, the

---
[59] 79 F. 984 (Ill. 1897).
[60] *Ex parte* Blois, 179 Cal. 291, 176 P. 449 (1918).
[61] Jefferson County Fiscal Court v. Jefferson County *ex rel* Grauman, 278 Ky. 785, 129 S.W.2d 554 (noted in 122 A.L.R. 1158, 1939).

court said the plan was within the city's power to promote the health, safety, and welfare of the public.[62]

Must a city go outside its boundaries to fight a fire? No. In fact, some states have statutes specifically permitting or prohibiting such a course of procedure. The more important question is, though, does the city have the power to send its apparatus outside its limits, and if so, what legal implications may arise as a result of undertaking such a practice?

To cope with area-wide disasters or conflagrations, whether of enemy origin or acts of God, many states have adopted laws which permit mutual aid agreements between various political subdivisions. In California the law authorizes use of city, city and county, or county fire protection equipment outside its territorial limits to extinguish a fire that is of such proportion that the individual unit cannot adequately handle it, and provides that a charge may be made for the reasonable value of the services rendered. It also gives the county power to contract with the city to assist the city in the prevention and suppression of forest fires, provides that a county may be reimbursed, and gives cities power to contract with each other with regard to fire equipment. A 1945 law allows the state to pay for all damages to equipment suffered while rendering outside aid during a state of emergency, and grants all the privileges and immunities from liability, pension benefits, workmen's compensation, exemptions from laws, etc., to persons responding outside their territorial limits to the same extent as if inside.

In sending its fire apparatus outside the city in response to a pay arrangement, the fire department's activity changes from a governmental one to proprietary, and hence renders the municipality liable for damage sustained by persons negligently injured by its fire trucks on such a call. It is immaterial that in the particular instance when the accident occurred the city received no remuneration if it was a general practice to charge for such services.

It might be argued, however, in favor of municipal nonliability that even though the capacity of the function in such cases changes from governmental to proprietary, in states holding that such a practice is *ultra vires*, the city would be immune. The City of San Angelo was not held liable by a Texas court for injuries sustained when its fire truck collided with an automobile where at the time of the collision the fire truck was on its way to extinguish a fire beyond the city limits under an agreement whereby the county furnished the truck and the city the men in fighting fires in the county. The court said: "If the city had no authority to answer calls to extinguish fires beyond the city limits, then its acts in doing so would be *ultra vires*, and the city would not be liable for the torts of its firemen while answering such calls."[63]

However, where a statute imposes municipal liability for the negligent operation of emergency vehicles, it should be immaterial whether the negligent act is committed within or outside the city limits, in considering the question of liability. The public policy which impelled the adoption of such a statute should compel such a conclusion. Note, however, in the closely analogous situation

---

[62] Burlington v. Industrial Commission, 195 Wis. 536, 218 N.W. 816.
[63] King v. San Angelo, 66 S.W.2d 418 (Tex. Civ. App. 1933).

where the negligent act was done while the municipal officer was acting outside the scope of his authority (in taking his wife to dinner in the police car), the California court interpreted the statute as not imposing municipal liability. Is a fire fighter acting outside the scope of his employment in fighting a fire across the city border?

A Colorado case[64] held that the authority to declare policy as to use of the city's fire equipment and fire fighters for extinguishment of fires outside the city limits rests with the city council, and until such action is taken the decision to do so or not to do so is discretionary with the city's responsible officials. After reviewing a number of cases in this field, the judge giving the majority opinion seemed to think that such outside assistance was justified from a humanitarian standpoint, and ruled that it could not be enjoined in a taxpayer's action. The court said:

"... The distinction, we think, is fundamental, and in cases involving actions of municipal authorities outside the city limits, it is important to distinguish which question may be involved: the power to enforce the authority of the municipality as against property or people outside its borders, or the right of municipal officers to act outside the city limits where no enforcement of authority is involved. The former right, if existing at all, is absolute; the latter is discretionary and dependent on purpose and result. The former, as raised in People v. Raims (20 Colo. 489), is dependent on legislative delegation of authority; the latter is dependent on whether or not the use of the city's officers or facilities or funds is, or is not, for the benefit of the municipality. Admittedly the operation of a fire department within the city limits is a proper municipal activity. Its purpose is to benefit the city, and, where there is no conflict of rights, its relation to that purpose, rather than arbitrary city boundary lines, must determine the extent of its use.

"As to the second ground, we cannot agree with the unsupported declaration of the trial court that the welfare and public interest of the municipality and the taxpayers therein are neither promoted nor protected by permitting the city's fire department to accept calls for the extinguishment of fire outside the municipality's corporate limits, even where public buildings are not involved. In many cases prompt action in extinguishing a small fire outside the city limits may prevent its increase and spread across the city line with disastrous results to the city and its taxpayers. The destruction by fire of a factory located outside the city limits may deprive resident taxpayers of means of livelihood and they and the city suffer loss thereby. Mutual assistance by neighboring cities may well work to the advantage of each. Perhaps in some cases the very good will acquired by assistance outside the city in emergency may be of value to the city and its taxpayers. On the other hand, there are doubtless many cases where large districts refuse to become annexed to the city, and the burden of furnishing fire protection to such areas becomes an imposition on the city and its taxpayers, and discretion might well require refusal of fire service except where there is danger of the fire spreading across the city lines...."

---

[64] City of Pueblo v. Flanders, 122 Colo. 571, 225 P.2d 832 (1950).

While such power exists under proper circumstances, a limitation against possible abuse of this power is found in the general requirement that any expenditure of city funds must be of benefit to the city or must improve the administration of its municipal affairs.[65] As stated in City of Pueblo v. Flanders, supra: "So we think the question here is not whether the actual service rendered by the fireman is within or without the City limits, but rather it is whether the predominant purpose is to benefit the City."

In a Connecticut case, the fact that a fire truck was being driven outside the city limits as a convenient route back to the station from the fire was held not to alter the old rule of municipal nonliability for the negligent operation of fire apparatus. In an Oklahoma case the city manager and trustee were held not liable to a civilian who had been "pressed into service" by the fire chief at a fire outside the city limits.

But what is the position of the fire fighter who is ordered to take apparatus outside the city limits to fight a fire? Is his liability greater? Is he still entitled to a disability pension in case of injury? Does his workmen's compensation still apply?

His liability and loss may be greater for an accident outside the city limits than for one inside. For example, a city ordinance may provide that fire apparatus has the right of way and for other vehicles to pull over to the curb upon hearing the siren. But if the state law contains no similar provisions, the general speed and traffic laws would govern the operation and conduct of a fire vehicle. Hence a greater degree of care would be imposed upon the driver of fire apparatus. His liability for negligence might then be considered the same as any other individual's.

A Massachusetts law was interpreted as making the reimbursement to a city of disability pension benefits to a fire fighter injured while aiding another city under a mutual aid plan, discretionary and not mandatory.[66] Though the question of fire fighters' pensions will not be considered at this point, it is interesting to note that his protection, not only for himself (in case of disability), but also for his family, may be jeopardized by reason of the fact that if a city is not empowered to fight fires outside its limits it would seem that a fire fighter would not be performing in the line of duty.

In a California case, however, the court said: "The fire chief was not acting outside the line of his duty because the fire was on the other side of the street which marked the city limits. He was under a duty to respond to the alarm whistle blown within the city limits. It was not shown that he was aware that the fire was beyond the city limits. Furthermore, protection of lives and property within the city required his presence at the scene of the fire, and the extinguishing of a fire scarcely over the line."[67] While this case is not authority for the proposition that a fire fighter is always acting in line of duty when responding to fires outside the city, it does indicate that under two situations, at least, it may be a question of fact that can be resolved in favor of such con-

---
[65] City of Roseville v. Tulley, 55 Cal. App.2d 601, 131 P.2d 395 (3rd Dist. 1942).
[66] City of Everett v. City of Revere, 344 Mass. 585, 183 N.E.2d 716 (1962).
[67] See note 53.

tention; viz., where it was not shown that he was not aware that the fire was outside the city, and where it was necessary to go outside in order to protect property inside the city.

Although the subject of workmen's compensation insurance is considered in a later chapter, it is relevant at this point to note that some states have adopted statutes which provide that all fire fighters, including volunteers, are eligible for such compensation while fighting a fire anywhere in the state. Under the applicable statutes of New York, for example, it was held that a volunteer fire fighter summoned by his own fire district to aid another district, and who was injured in the line of duty, was entitled to compensation by the district that was being aided.[68]

Is the city liable to the owner of a building within the city if his property burns while the fire department is outside the city fighting a fire? Should there be a distinction, as far as the municipality's nonliability is concerned, between its negligently doing an authorized function and its failure to perform the function at all? While there are no cases directly in point, it is probable that no recovery could be had. Irvine v. Chattanooga[69] furnishes an interesting illustration of how inattentive a fire department can be without incurring liability to the city. Picture a small fire in the roof near the chimney of the plaintiff's home. After failing to get any response to an alarm box signal to a station only two blocks away, the plaintiff frantically turned in a general alarm to every station in the city—but still no response. Frantically he tried to reach the department by telephone, but without success. Meantime the fire had spread to other portions of the building, and notwithstanding the plaintiff's utmost exertions his house was totally destroyed. It was reasonably certain that if the fire department had responded it could easily have extinguished the incipient fire, but the whole department was on the opposite side of the city—two miles away—on parade duty while attending the funeral of the city physician. The court announced that there was no implied contract to extinguish every fire in the city merely because the fire department was supported by taxes, any more than there was a contract to catch every criminal.

In Yule v. New Orleans[70] nearly all the fire fighters were at the fairgrounds, some miles distant from their engine houses, where by law they should have been, and "spent the day in feasting, fun and frolic," when quite a number of dwellings burned down. In holding the city not liable the court explained that it was not an insurer against fire losses.

Mutual Aid and Assistance

The general rule is that a fire department may not operate outside its jurisdiction unless expressly authorized to do so by statute. Several cases have held that, although it is proper to obtain fire protection from outside aid, a city

---
[68] Gilewski v. Mastic Beach Fire District No. 1, 199 App. Div.2d 299, 241 N.Y.S.2d 874 (1963).
[69] 101 Tenn. 291, 44 S.W. 419 (1898).
[70] 25 La. Ann. 394, 9 A.L.R. 144 (1873).

has no authority to enter into a contract to help another jurisdiction unless an express law so allows. Many states have laws which permit governmental entities to enter into mutual aid agreements to help each other in case of great peril, extreme emergency, or public calamity. What is needed is a law to enable fire departments to assist each other on a day-to-day basis, and to cover-in for the adjacent town while its apparatus is out on a fire or rendering aid to its neighbor.

There is no question, however, but that a fire department may go outside its city limits to fight a fire which may come into the city, law or no law. Whether a fire chief runs the risk of a law suit for the negligent acts of fire fighters from another department who come in and assist him with his fire is another question. This last point is covered in the California Government Code, for example, and the assisted agency is not liable under such circumstances; the law also permits the towns to enter into a contract to reimburse the assisting city in such a case.

However, not many of these contractual arrangements have been entered into in the state California. Every city apparently prefers to bear its own liability—to pay for its own judgments and to ask for no reimbursement of any kind, whether the reimbursement be for damaged equipment, lost hose and fittings, or anything else.

In Woodworth v. Village of Watkins Glen[71] there was no liability imposed upon a village which was holding a scheduled drill of volunteer fire companies from several other municipalities for injuries to a member of one of the visiting companies for it had not called for assistance within the meaning of the workmen's compensation law. Two other cases delineating liability for injuries and death are:

"Under such a statute, liability has been imposed for damages to a pumper and its accessories, but denied with respect to injuries sustained by firemen."[72]

"But a fire force that remains under the direction and control of its own chief during an extraterritorial engagement remains in the employ of its own municipality which will be liable under Workmen's Compensation laws for the death or injury of any of its firemen while so engaged."[73]

The argument has been used in the past that where a city has no authority to perform any service beyond its boundaries, then it doesn't matter how carelessly such acts are conducted, it could not be held liable. But such authority (to go outside the limits) may be implied, as in a situation where it is necessary to help a neighbor stop a fire so that it doesn't spread into the city, or such authority may be granted by statute.

For an example of the latter instance, there is a North Carolina case which declared that acts performed by municipal agents outside the corporate limits are not necessarily beyond the scope of their authority—the exact wording was "... not *ipso facto ultra vires*"—and hence a city may be liable for torts committed beyond its boundaries if the elements of liability are present.

---
[71] 7 App. Div.2d 694, 179 N.Y.S.2d 226 (1958).
[72] City of Watertown v. Town of Watertown, 207 Misc. 433, 139 N.Y.S.2d 198 (Sup. Ct. Spec. Term 1952).
[73] Nelson v. Borough of Greenville, 181 Pa. Super. 488, 124 A.2d 675 (1956).

Problems In Fire Extinguishment
Failure to Extinguish Fires

At this point it might be well to consider the general question of municipal liability for failure to extinguish fires, or for failure to supply sufficient water, hydrants, or hose.

If a city is under no obligation to provide fire protection in the first place, it logically follows that it is not liable for failing to extinguish a fire when it has organized a fire department, unless one could make out a case of promissory estoppel because of reliance on the city to furnish fire protection. Such an attempt would have been futile, however, under the old rule that fighting fires is a governmental function for which there is no municipal liability attendant upon its nonperformance. A similar result (nonliability) can be reached, however, without invoking this archaic doctrine. In Steitz v. Beacon,[74] a case decided by the New York court of appeals, an action was brought against the City of Beacon to recover damages suffered as the result of a fire. The complaint alleged that the plaintiff's property was destroyed because of carelessness and negligence by the city in failing to create and maintain a properly-equipped fire department for the benefit of plaintiff's property and that the city negligently failed to maintain water pressure in mains near the plaintiff's property. In dismissing the complaint the court held that the city could not hide behind the cloak of sovereign immunity but that likewise there was no liability known to the law by virtue of which it was the city's duty to quench the fire or indemnify the loss. The responsibility of the city as far as fire was concerned under its general grant of power, the court ruled, was to protect the public health and safety and to provide for the general welfare. Further, the responsibility was not designed to protect the personal interest of any individual but was for the benefit of all members of the community; that although there was indeed a public duty to maintain a fire department there was no suggestion that the people of the city could recover fire damages to their property for any omission in keeping hydrants, valves, or pipes in repair.

If, as in Irvine v. Chattanooga[75] and Yule v. New Orleans,[76] instances where the failure to extinguish the fire was because fire fighters were out on parade or at a picnic, the city was not liable; the same is true where fire fighters cannot open the hydrant with their spanners and two houses burn down while they look for a monkey wrench,[77] or because fire fighters cannot get within fighting range of the fire because of the city's failure to provide enough hose (or a hydrant close enough to the fire).[78]

Robinson v. Evansville[79] furnishes a good example of what difficulties may be encountered in extinguishing a simple fire. The plaintiff's house was only

---
[74] 295 N.Y. 51, 64 N.E.2d 704 (1945).
[75] See note 69.
[76] See note 70.
[77] Wright v. Augusta, 78 Ga. 241, 6 Am. S.R. 256 (1886).
[78] Small v. Frankfort, 203 Ky. 188, 201 S.W. 1111 (1924); Larrimore v. Indianapolis, 197 Ind. 457, 151 N.E. 333 (1927).
[79] 87 Ind. 334, 44 Am. Rep. 770 (1882).

700 feet from the nearest hydrant, and the fire station was only 600 feet from the hydrant, but the hose rig only carried 650 feet of hose (though its capacity was 800 feet) and hence would not reach the fire. While waiting a half hour for another hose company to arrive (due to a balky horse), the fire fighters were endeavoring to get water from the hydrant that was full of mud. When the hose was finally connected, it broke in pieces under the pressure. Needless to say, the plaintiff's house was a total loss, but the city was not held liable.

Rekindles

Significance of the "out tap" sounded by fire departments to denote that a fire has been extinguished was taken under consideration by the New York State Court of Appeals in a case that arose from a fire that was put out in the plaintiff's home. The chief ordered the traditional out tap sounded when it seemed the fire was out. But the plaintiff noticed what appeared to be smoke and he mentioned it to the chief. The chief was quoted as saying, "It's only vapor." But after the fire fighters left the fire broke out anew and the house was totally destroyed.

In arguments before the high court, the plaintiff's lawyer contended that it isn't so much a question of negligence of fire fighters in putting out a fire, as it is the judgment of an expert, the fire chief, that the fire is out—as signified by the out tap. Supreme Court Justice Willard Best dismissed the action on the grounds that municipalities, by decisional law, are not responsible for damages resulting from "governmental acts" such as by fire departments.

Failure to Provide Sufficient Water for Fire Protection

Does a property owner have a right to recover for a loss by fire due to an inadequate water supply? The answer to this question depends not only upon the jurisdiction, but upon whether the plaintiff sues the water company or the city, and whether the water company is privately owned, city controlled (by stock ownership), or municipally owned.

The courts in a large majority of the states, including Alabama, California, Connecticut, Idaho, Illinois, Indiana, Iowa, Ohio, Oklahoma, Pennsylvania, South Carolina, Tennessee, Texas, West Virginia, and Wisconsin, as well as the Supreme Court of the United States, hold that a private property owner cannot recover against a private water company whose contract is with a city for loss by fire on account of the company's negligence. Lack of privity of contract between the property owner and the company precludes recovery.[80]

The Supreme Court of the United States, in a case involving a water company that had a contract with a city to furnish adequate fire protection, said that whether the failure to provide the protection be regarded as *ex contractu* or *ex delicto*, it was a breach of the contract the company had with the city, and a majority of American courts do not allow a private citizen to sue for such a breach of contract. The Court further added that no tort liability exists be-

---

[80] See cases collected in 3 GEO. WASH. L. REV. 270 (1935), "Municipal Corporations: Right of Property Owner to Sue for Loss Because of Inadequate Water Supply."

cause the defendant owed no duty to the plaintiff to furnish him with fire protection, for the city owed him no duty to do so, and the defendant's duty extended only to the city. The Court went on to say:

"In order for a third party to sue on a contract, in jurisdictions where the right is granted, he must show that it was intended for his direct benefit. A private citizen is interested in the faithful performance of contracts of service by policemen, firemen, . . . as well as in holding the vendors of fire engines to their warranties. . . . But for breaches of their contracts the citizen cannot sue though he suffer loss because . . . the firemen delayed in responding to an alarm, or the engine proved defective, resulting in his building being destroyed by fire."[81]

Kentucky, however, along with Pennsylvania, Florida, and North Carolina, has followed the minority view, and in Paducah Lumber Co. v. Paducah Water Supply Co., the court said: "A party for whose benefit a contract is evidently made may sue thereon in his own name, though the engagement be not directly to or with him . . . ," and allowed a recovery against the defaulting water company which had a contract with the city.[82]

On the other hand, if a city owns all of the stock in a private waterworks, or owns the waterworks, the Kentucky courts hold that the city is not liable to individual citizens for damages resulting from failure to supply sufficient water for fire protection. But a New York case[83] held that a city empowered by its charter to construct and operate a waterworks and to maintain a fire department is not liable for the destruction of property by fire because of negligent failure to keep in repair and properly to operate certain pressure regulating valves in its water system, by reason of which an insufficient quantity of water was provided to combat a fire effectively. The annotation in 163 ALR 348 discusses "Liability of municipality for fire loss due to its failure to provide or maintain adequate water supply or pressure."

While it may be liable in its proprietary capacity to its citizens for failure to supply sufficient water for domestic use, the majority of cases hold that a municipality is not liable for negligence in failure to supply adequate water for extinguishment of fires. The courts differ, however, in their reasons for such a holding. In one case the liability was barred on grounds of sound public policy, in another because the failure to repair the cistern was too remote to be the proximate cause of the fire loss, and in others because the furnishing of water for fighting fires is a governmental function.

Those states holding the municipality liable do so on the grounds that the business of selling water to the public is a corporate function. According to the Federal Court in United States v. Sault Ste. Marie,[84] the city had no power to contract liability for failure to maintain a minimum pressure and hence the United States was not entitled to recover a fire loss resulting from the breach of such a contract.

---
[81] German Alliance Insurance Co. v. Howard, 226 U.S. 220, 33 Sup. Ct. 32, 42 L.R.A. (n.s.) 1000 (1912) at p. 1004.
[82] Paducah Lumber Co. v. Paducah Water Supply Co., 89 Ky. 340, 12 S.W. 554 (1889).
[83] See note 74.
[84] 137 Fed. 258 (Mich. 1905).

Failure to Inspect for Closed Valves

Because a valve was closed in a water main which supplied a fire hydrant in the water system operated by the City of Grand Coulee, all the plaintiff's buildings were destroyed by fire, including a laundry, restaurant, and tavern. The Court of Appeals of Washington, in granting a judgment for the plaintiff, held that a city which maintains a water system to which fire hydrants are attached has a duty to regularly inspect that system to ensure an adequate water supply to those hydrants; Failure to inspect is a breach of duty as a matter of law.[85] entity if it were *a private person.*' (Italics added.)

In states having laws which define the liability of government agencies by statute, such as California, the courts have found a basis for nonliability on statutory interpretation. In the case of Heieck and Moran v. City of Modesto[86] the plaintiff alleged that the fire department was notified of a fire that had started at premises in the city adjoining the premises of the plaintiff; that the fire department promptly responded with sufficient personnel, equipment, and facilities to contain the fire, but, because city employees had closed a valve in the water main, there was no water in the fire hydrants; consequently the fire spread to the plaintiff's premises. It was further alleged that the valve had been closed to permit relocation of certain water mains, but although the relocation had been completed at least a month before the date of the fire, the valve had not been turned on; and that no city employee notified either the city fire department or plaintiff that the water was shut off.

It was also alleged that the city "well knew the County of Stanislaus maintained a fire department, with tank trucks, which was ready, willing, and able to respond to said fire . . . (but defendant city) failed and neglected to notify said county fire department of the existence of said fire and negligently and carelessly failed to request its assistance in containing and extinguishing same."

In holding that the city was not liable for the water failure the court said:

"First, section 815 of the Government Code declares that 'Except as otherwise provided by statute: (a) A public entity is not liable for an injury, whether such injury arises out of an act or omission of the public entity or a public employee or any other person.

"(b) The liability of a public entity established by this part (commencing with Section 814) is subject to any immunity of the public entity provided by statute, . . . and is *subject to any defenses* that would be *available to* the public entity if it were *a private person.* (Italics added.)

"In Stang v. City of Mill Valley (1952) it was held that the Public Liability Act of 1923 did not impose liability for a fire loss to property due to defective firefighting equipment (there, water lines clogged with refuse). The ground of the holding was that a private water company would not be liable, which in turn was based upon earlier cases holding that neither a city (Ukiah v. Ukiah Water and Imp. Co. (1904), nor a private citizen (Niehaus Bros. Co. v. Contra Costa Water Co. (1911), could recover damages from a water company for a

---

[85] Morse, H. Newcomb, "Legal Insight: Dry Hydrant Costs City," *Fire Command!,* Vol. 41, No. 5, May 1974, p. 28.
[86] Heieck and Moran v. City of Modesto, 64 Cal.2d 229, 49 *Cal. Rptr.* 377, 411 P.2d 105 (1966).

property loss by fire due to the company's failure to maintain its water system properly. It appears that in Ukiah the complaint was of insufficient water pressure at the fire hydrants, and in Niehaus the court found that (as alleged here) defendant had no 'water supply in its mains available at the hydrants.' Thus Stang did not turn on the doctrine of sovereign immunity, it was not overruled by Muskopf, and it is now reflected in the quoted provisions of section 815, subdivision (b), which confirm the absence of liability of the city under the circumstances here alleged by plaintiff. This holding likewise disposes of plaintiff's contention that liability is cast on defendant by the provisions of subdivision (a) of section 815.2 of the Government Code, which covers only acts for which a governmental employee would be liable.

"Second, sections 850.2 and 850.4 of the Government Code expressly give immunity in this case. Section 850.2 provides that 'Neither a public entity that has undertaken to provide fire protection service, nor an employee of such a public entity, is liable for any injury resulting from the failure to provide or maintain sufficient personnel, equipment or other fire protection facilities.' Section 850.4 states that 'Neither a public entity, nor a public employee acting in the scope of his employment, is liable for any injury resulting from the condition of fire protection or firefighting equipment or facilities or (with an exception relating to motor vehicles) for any injury caused in fighting fires.' The complaint in the present case alleges that city employees while acting in the scope of their employment closed a water valve and left it closed. Thus whether the alleged injury to plaintiff's premises be viewed as resulting from 'failure to provide or maintain sufficient . . . fire protection facilities' (ss 850.2), or from the closed 'condition' of the water valve (ss 850.4) the conclusion is inescapable that the Legislature intended to establish immunity under the circumstances alleged by plaintiff.

"Citing Morgan v. County of Yuba (1964), plaintiff argues that failure of some member of defendant city's fire department to summon the county fire department was actionable. However, in Morgan it was the failure to carry out an alleged earlier promise to give warning (of release on bail of a dangerous prisoner) which was held to constitute one element of the asserted cause of action, rather than the making or the decision to make the promise. In the present case it is not alleged that any city employee promised or represented that other assistance would be summoned when lack of water in the mains was discovered, and no duty to summon such assistance is either suggested or shown by plaintiff."

Failure to Provide Automatic Sprinklers or Other Fire Protection

Ten years after the preceding case was decided, a California Appellate Court was called upon to construe Government Code Sections 850 and 850.2 in the case of Vedder v. County of Imperial.[87] The court declared:

"The statutes must be strictly construed, and governmental immunity should

---

[87] 36 Cal. App.3rd 654 at 660, 111 *Cal. Rptr.* 728 at 731 (4th Dis. 1974).

not be decreed unless the Legislature has clearly provided for it. They should not be applied to allow a public entity to escape responsibility for damages resulting from its failure to provide fire protection in its own property which it owns and manages itself, particularly where it has permitted a dangerous fire condition to exist on the property. In that situation, lack of fire protection is a proper factor to be considered as contributing to the existence of a dangerous condition on the property."

Fire Hydrants

Closely related to the question of failure to furnish adequate water supply are the problems arising out of municipal ownership of fire hydrants. In some states, where an injury was caused not from a fire but from a city's negligence in the installation or maintenance of its fire hydrants, the municipality has been held liable even though such hydrants are used for governmental purposes. States making a municipality liable for injuries in connection with fire hydrants are California, Connecticut, Georgia, Indiana, New York, and Missouri; decisions holding the municipality not liable are found in Minnesota, New Hampshire, North Carolina, and Vermont.

On the other hand, where a loss was sustained as a result of a fire department using a hydrant in extinguishing fires, as for example where water ran from a hydrant during the course of fire pumping operations into adjacent cellars, the city was not liable, even though such hydrants were also used on other occasions for corporate purposes, such as flushing streets, or watering trees. In the absence of a court decision or statute, a city is not liable for the negligence of its fire fighters when performing a governmental function, and fire fighting is a governmental function.

Nor was a city liable where a fire fighter thawed ice in a hydrant by letting water run onto the street resulting in injury to a pedestrian who slipped, nor where a hydrant was located in a street so as to menace traffic. In both instances the hydrants were separate from the domestic water supply and were used exclusively for public purposes.

Where the duty has been imposed upon the fire department to test and flush the hydrants, this has been held to be within the proper scope of fire department functions, governmental in nature, and hence the municipality was not liable for injuries sustained as a result thereof.

Nor was a municipality liable for property destroyed while the fire department took an hour getting the stones, sand, and bark out of a hydrant, providing the municipality made no charge for any of its services connected with its water system.

Neither was a municipality liable for a house destroyed while the fire fighters searched for a hydrant hidden under a pile of paving materials workmen had deposited over it in violation of an ordinance. There was a statute providing that in no case shall the municipality be liable for a failure to furnish an adequate water supply.

Some municipalities exact, as a condition to supplying water service, a written waiver of liability for any damage resulting from a failure to supply sufficient water or pressure, or from excessive pressure; nevertheless, liability may be imposed for injuries resulting from negligence in installing and maintaining fire hydrants where the damage inflicted is something other than a loss by fire.

In a case where water from a city-owned water works shot out of a fire hydrant, breaking windows and filling basements across the street, a Missouri court, while disapproving an earlier case holding the city liable on the corporate function theory for leaky hydrant damage to a cellar, held that the city might be liable for creating a nuisance and submitted a *prima facie* case to the jury under the *res ipsa loquitar doctrine*.

Municipalities have been held liable in other cases where reasonable care in making inspections would have revealed defects in hydrants or the pipes connecting them to the water mains, as where the lateral water pipes break and flood a cellar, or where the water percolates through the soil causing a building to settle and crack. And this is true even where a pipe which has flooded a cellar had burst because of added pressure thrown on the system through pumping operations during a fire. Such extra pressure should have been anticipated.

A municipality is also liable to pedestrians who injure themselves on unguarded fire hydrants that obstruct the sidewalk. In California, it was also no defense to a municipality that a householder failed to keep his roof hatches in place, and when water from a city main shot up in the air it flooded his building from the roof.

Fire Hydrant Rental and Related Water Charges

Even where fire departments do most of the inspection, testing, and painting of fire hydrants, the departments are often required to include in their annual budget a sum sufficient to pay the monthly rental fee charged by the water company, whether publicly or privately owned. This may be the result of court decisions or state statutes.

In California, for example, it is both. Prior to the passage of Section 53069.9 of the Government Code of California, which authorizes any public agency providing water for fire protection purposes to repair and to collect a charge to pay the costs of installing and maintaining fire hydrants, the court decided in Arcade County Water District v. Arcade Fire District[88] that the fire district was liable for the payment of water charges assessed by the water district.

A 1974 decision which relied upon Section 53069.9 of the code granted the Northridge Water District a judgment against the Citrus Heights Fire District, although it is not clear from the provisions of the statute that such a judgment can be satisfied out of property belonging to the fire district, since it provides that such charges may be made on all land within the public agency to which water is made available for fire protection purposes.[89]

---

[88] Arcade County Water District v. Arcade Fire District, App. 85 *Cal. Rptr.* 737 (1970).
[89] Case No. 171032, Sup. Ct., Sacramento County.

Torts of Fire Fighters

We have seen that municipalities are not liable for the retention of incompetent fire fighters, nor for the torts of fire fighters when performing a governmental function, or when they act outside the scope of their authority, or in jurisdictions holding that fire fighters are public officers instead of employees (and hence no *respondeat superior*). Yet municipalities are liable for the torts of their fire fighters where statutes provide for it, or where their acts create a nuisance, or when they are engaging in a nongovernmental function.

While many cases hold that a municipality is not liable for the negligence of its fire fighters "while engaged in the discharge of their duties," the qualification is superfluous unless it means "governmental duties." A municipality per se is not liable for acts of its fire fighters who act outside the scope of their duties, e.g., who without authority start a fire to clean off a vacant lot, while it is liable where the duties of the fire fighters constitute a proprietary function, or create a nuisance.

Other Fire Equipment

With respect to municipal liability, the following discussion deals with injuries arising from negligent maintenance of fire equipment other than fire engines, fire boats, or fire hydrants.

Connecticut courts have held that testing a fire whistle is a governmental function; hence, there is no municipal liability to one who is injured when the compressed air vessel ruptures.[90] Since horses are still used by some large municipal departments in connection with their patrol stations, it is relevant to note that there exists a municipal immunity for the trespass of horses used exclusively by the fire department.[91]

In Hammond v. Atlanta,[92] where the city was held not liable for the negligence of a fire fighter in knocking down a pedestrian with a hose reel, the court distinguishes Augusta v. Cleveland,[93] where it was held that the creation of an obstruction of the streets by employees rendered the city liable, even though they were engaged in a governmental function.

There are several cases that have held a municipality not liable to pedestrians who stumbled over fire hose, even though such negligence was predicated upon the failure to maintain the streets and sidewalks in a reasonably safe condition. "If this necessary fire fighting apparatus crossing a sidewalk renders travel dangerous, temporary safety must yield to the more important and imminent demand for saving property and perhaps life by promptly suppressing the conflagration."[94] This immunity would probably not exist, however, in a case where the hose stretched across the sidewalk was not being used in connection

---

[90] O'Donnell v. Groton, 108 Conn. 622, 144A (1928).
[91] Cunningham v. Seattle, 40 Wash. 59, 82 P. 143 (1905).
[92] 25 Ga. App. 259, 103 S.E. 39 (1921).
[93] 148 Ga. App. 734, 98 S.E. 345 (1919).
[94] Powell v. Village of Fenton, 240 Mich. 94, 214 N.W. 968 (1927).

with fire extinguishing operations, as for example, where it is stretched out on the sidewalk in front of the station to be washed or dried.

In an early case the Massachusetts court held that the City of Boston was not liable to one who was injured by the bursting of a fire hose at a fire, because the fire fighters were said to be officers and not agents of the city.[95]

Where an unusual windstorm blew over the improperly constructed fire alarm tower (with an 800-pound bell in it), damaging the property of an adjoining owner,[96] and where the wall of a fire cistern collapsed on an employee engaged in shoveling out silt,[97] municipal liability was imposed.

Fire alarm signal wires have also been a source of liability to various cities. Where a pole used solely for the support of a fire alarm box and signal wire broke off at the ground where it had rotted through, falling on a boy playing in the street, a city was held liable, for the following reasons:

"It is the duty of the city to exercise reasonable care to keep its streets free from danger, and it possesses the power to do whatever is necessary to accomplish this purpose. If, through negligence, it fails to perform this duty and an injury results from a dangerous condition of one of its streets, the city cannot avoid liability by showing that negligence in caring for an instrumentality used in the performance of a governmental duty is what brought about the dangerous condition. . . ."[98]

The same state, eleven years later, held that a man who had volunteered to help the fire department replace a fire alarm wire which was broken in the course of trimming a tree, could not recover from the city for injuries sustained when the pole against which his ladder leaned broke and caused him to fall into the street. In explaining how the previous decision differed and did not apply the court said:

"Hilstrom v. City of St. Paul is not a departure from the rule . . . (against municipal liability for negligence in performance of governmental functions), for there the liability arose, not because the city had negligently failed to maintain its fire alarms' system in a reasonably safe condition, but because of its negligent failure to inspect the pole, it created a street hazard and made the street dangerous for public use."

A more reasonable ground for distinction lies in the fact that the injured party was not a mere pedestrian, but one who had voluntarily assumed the ordinary risk incident to repairing damage that he had done to the signal wire.

Fire Station and Premises

Since fire stations are often visited by children on school tours, students of civic and social sciences, fire fighters from other cities, fire fans, and others out of mere curiosity, it may be well to consider the city's responsibility to these persons for injuries received through defects in the buildings or premises.

---

[95] Fisher v. Boston, 104 Mass. 87, 6 Am. Rep. 196 (1870).
[96] Wilson v. City of Mason, 190 Ill. App. 510 (1914).
[97] Mulcairns v. Jonesville, 67 Wis. 24, 29 N.W. 565 (1886).
[98] Hilstrom v. City of St. Paul, 134 Minn. 391, 214 N.W. 656 (1927).

Extra precautions should be taken in conducting tours for children. For example, when a fire captain in the Central Fire Station in Baton Rouge, La., climbed into the cab of a fire truck to demonstrate the siren, he warned everyone to step back, but he could not see a little nine year old girl stick her finger in the siren. When he pushed the button, the siren's blade amputated part of her finger. Since the child's mother, aunt, and seven cousins were present at the time, and since the captain's testimony that he had given them all warning to stand clear was believed by the trial court, the city was not held liable and the decision was affirmed by the Court of Appeals of Louisiana.[99]

With respect to pole hole injuries, in jurisdictions holding that the maintenance of a fire station is a governmental function, there is no municipal liability for injuries to members of the public who fall through such openings. Thus in Nicastro v. Chicago[100] where a child, taken into the station by a fire fighter without authority to do so, fell through the pole hole, the fire fighter, but not the city, was held liable, for it was the negligence of the fire fighter and not a defect in the building that caused the injury. Also, in Barnes v. Waco,[101] where a doctor, who was called to examine a fire fighter applicant for an insurance policy, fell through a pole hole and was killed, the court held that he was a mere licensee to whom the city owed no duty to keep the property in a safe condition. Therefore, despite dictum to the contrary, these cases are not authority for the proposition that a city is not liable to an invitee who is injured through defects in a fire station. In the case of Abihider v. Springfield,[102] the nonliability of the city was based upon the fact that the pedestrian was injured upon the sidewalk approach to the station, rather than on the public way. Nor does Adkinson v. Port Arthur[103] refute municipal liability to an invitee, for in this case the court held that a fire fighter could not recover under the workmen's compensation law for injuries received at a defective drill tower.

On the other hand, it has been positively asserted that in the care and management of a fire station a city is performing a ministerial duty, and is liable for injuries caused by defects of which it has notice and reasonable opportunity to repair. Although not expressly using the term "estoppel" as the basis for its decision, the Kansas court in effect adopts such an argument when it said:

"When a municipal corporation assumes the performance of a public duty which was permissive only and enters upon the discharge of such duty, and through the negligent performance thereof by its authorized agents one is injured either in person or property, the corporation cannot escape liability by saying that the performance of the duty was not imperative."[104]

In holding the city liable for injuries inflicted upon a passerby when the station doors sprung open after being released, the Kansas court charged:

"If the operation of these doors with reasonable care would have provided

---

[99] Morse, H. Newcomb, "Legal Insight: Accident in a Fire Station—Negligence?," *Fire Command!*, Vol. 40, No. 5, May 1973, pp. 48, 49.

[100] 175 Ill. App. 634 (1912).

[101] 262 S.W. 1081 (Tex. Comm. of App. 1924).

[102] 277 Mass. 125, 177 N.E. 818 (1931).

[103] 293 S.W. 191 (Tex. Cir. App. 1927). Nor can fire fighters recover damages from city for pole hole injuries, according to Miller v. Macon, 152 Ga. 648, 110 S.E. 873 (1922).

[104] Bowden v. Kansas City, 69 Kan. 587 at p. 594, 77 p. 573, 66 L.R.A. 181, 105 Am. St. Rep. 187 (1904).

against danger and accident to the passerby the city is not liable. If the necessary and natural and probable operation of these doors was dangerous even though accompanied by the use of ordinary care on the part of the employees, the city is liable for the result."[105]

The same result was reached in Walters v. City of Carthage[106] where a station door fell on a three-year-old boy playing on the sidewalk. The court held, "A municipal corporation is performing a ministerial duty in maintaining a fire station and is liable in damages for neglecting to make the same safe."

Is a Fire Station an Attractive Nuisance?

Is a fire station a nuisance? Not per se, but before proceeding to a discussion of the application of the attractive nuisance doctrine, it might be well to note in passing that a court may grant an injunction against the erection of a fire station at the insistence of abutters. It is a discretionary matter to do so pending a hearing on the merits as to whether or not a station would constitute a nuisance under the circumstances.

Particular care should be exercised with respect to children who are frequent callers at fire stations, and who sometimes slip by the fire fighters on floor watch, unnoticed, to wander bewilderingly around the apparatus floor. It may be no defense to the city that the injured child was a trespasser if the court holds that it is a proper case for the application of the attractive nuisance doctrine. The Supreme Court seems to make the liability rest upon the allurement of the nuisance, as well as the defendant's anticipation of danger to children. The elements necessary to apply the doctrine are:
1. A condition novel in character.
2. Attractive and dangerous to children.
3. Easily guarded and made safe.

Certainly fire apparatus is no less attractive to children than railway turntables, roofing kettles, pools or lumber piles. Hence the fire fighter on floor watch should be particularly careful to safeguard children from possible injury at this source. Under the attractive nuisance doctrine, a child who is tempted to play about dangerous machinery because of its innate fascination for children is treated by the law as an invitee rather than a trespasser, raising the duty to use due care to prevent the child from being injured.

In Melendez v. City of Los Angeles[107] the court seemed to indicate that the attractive nuisance doctrine was applicable to actions under the California Public Liability Act, and although the plaintiff was not permitted to recover for the drowning of two children, it was on the grounds that the particular pool was not an attractive nuisance.

Statutory Liability for Dangerous Conditions

In some jurisdictions liability may be imposed by statute on a municipality for injuries caused by defects in a fire station and its premises. Cases interpret-

---
[105] Kies v. Erie, 169 Pa. St. 598, 32 A. 621 (1895).
[106] 36 S.D. 11, 153 N.W. 881 (1915).
[107] 8 Cal.2d 741, 68 P.2d 971 (1937).

ing the California statute on liability illustrate the problems involved. The California legislature has imposed liability on cities, counties, and school districts for injuries to persons and property or death resulting from the dangerous or defective condition of public streets, buildings, grounds, works, and property, where the council or other officer having authority to remedy such dangerous or defective condition had knowledge or notice thereof, and yet failed or neglected to remedy the same, or to protect the public against it within a reasonable time thereafter.

It has been urged that the public property referred to in the act is that which has been set aside for use by members of the public and does not include property on which members of the public would be trespassers. It is not likely that the legislature intended to give members of the public a right of action for injuries arising out of a dangerous condition on public property in places where they have no right to go.

Hence, a visitor to a fire station with permission to view the apparatus on the ground floor should not be able to recover for injuries sustained in the dormitory, nor should he be able to recover for damages received by reason of a hose of falling on him in the hose tower, where he was not invited.

However, in order for a California resident to determine a good cause of action, the essential elements of a cause of action under the California act are:

1. A dangerous condition of the property.
2. Actual or constructive notice by persons having authority to remedy the same.
3. The lapse of a reasonable time after notice to remedy such condition.
4. The dangerous condition must be shown to have been the proximate cause.
5. And a verified claim must have been presented within the statutory period.

While the act says that a city shall be liable for injuries resulting from "dangerous or defective conditions," the court has said that the "defective condition" must also be "dangerous," and unless it is of such a minor character that it could not be dangerous as a matter of law, the jury determines whether it is dangerous or not.

A dangerous condition may arise from improper construction or a failure to maintain after construction. It may result from the absence of warning devices and guard rails; the presence of slippery steps, holes in pavement, cracks in sidewalks; or the failure to make minor repairs.

Liability under the act may be imposed irrespective of whether the party injured is a licensee or invitee. An example is where a young lady, about to cut through a county court house for convenience, was forced off a slippery, worn metal door step by a self-closing door, when she turned around to watch fire engines (the siren having attracted her attention). In holding the county liable, the court said:

"In connection with liability imposed (speaking of the act) no exemption therefrom is expressed where the asserted defects are patent, nor is any distinction indicated as between a licensee or an invitee who may sustain injuries by reason of a defective or dangerous condition of the public property mentioned in the act."[108]

In Boothly v. Town of Yreka City,[109] the city had let the county use gratuitously one of the upstairs rooms of the engine house for a polling place. The plaintiff's heel caught on a nail that was sticking up on a poorly lighted stairway, causing her to fall. In holding the city and county liable, the court said that the city owed a duty to the plaintiff, as a third party, to see that the premises were reasonably safe. The court added that insofar as the fire department had knowledge of the defect, both the landlord and tenant are liable. Whether or not the tenancy was for compensation was immaterial.

The California courts have interpreted the notice required of the act to include constructive notice, and have extended it so as to require reasonable inspection. "The existence of a conspicuous defect or dangerous condition in a sidewalk or street which had existed for a considerable length of time (three months) will create a presumption of constructive notice thereof."[110]

The political subdivision is allowed a reasonable time in which to repair the defect after acquiring notice of its dangerous character, and five or six days have been held sufficient for minor repairs.

After it is established that the defect was the proximate cause of the injury, it is no defense to the city that there may have been a concurring cause, but contributory negligence will release the governmental unit from liability.

Where there is a statute similar to the California act just discussed, it should make no difference, as far as liability is concerned, whether the injury occurs in a public building used for the performance of governmental or ministerial functions. In view of the decisions holding that the maintenance of fire stations is a ministerial function, it would seem that such buildings and premises should be maintained in a safe condition, even in the absence of a statute, if municipal liability were to be averted for injuries resulting from defects therein.

Thus it is advisable for every fire fighter, upon discovering any condition that renders the use of the building or premises unsafe, regardless of how minor it may seem at the time, to report it immediately to his superior officer. Such notice will not only give the city a chance to make repairs, but it will be a material element in any cause of action that may accrue in the event that a visiting friend or even a fellow fire fighter suffers any injury.

Destruction of Buildings to Prevent the Spread of Fire

At common law a municipality needs no statutory authorization to destroy buildings to prevent the spread of fire, and is not liable for the destruction of either a building or its contents under such circumstances. And the power to destroy buildings carries with it the power to determine the necessity of so

---
[108] Gibson v. Mendocino County, 16 Cal.2d 80, 105 P.2d 62 (1940).
[109] 117 Cal. App. 643, 4 P.2d 589 (1931).
[110] Adams v. So. Pac. Co., 4 Cal.2d 731, 53 P.2d 121 (1935).

doing. There are different grounds asserted for the right to exercise this power, such as public welfare, private necessity, police power, and abatement of a nuisance; i.e., the house on fire, eminent domain, and statute. In the latter two cases, liability is imposed; in the former because the taking under eminent domain carries with it the correlative duty to pay just compensation, and in the latter because statutes usually provide that the person whose house is destroyed may recover under certain conditions. These conditions should be carefully complied with, however, for the right is given in derogation to the common law, and will be strictly construed.

In a Boston case, recovery was denied because the facts showed that the fire chief had given his men blanket authority to pull down buildings, or blow them up, either removing the contents or not, at their own discretion, whereas the statute required as a condition to municipal liability that the order to destroy be concurred with by three fire engineers.

Riots—Participation in Civil Disturbances

Fire departments throughout the world are sometimes called upon to assist law enforcement agencies in quelling riots and public disorders. Sometimes there are no fires involved, and oftentimes the fire chief has some serious reservations as to whether or not he should tie up his equipment on one side of the town for mob control when at any moment a fire might break out on the other side of town, either accidentally or deliberately. He has a right to be concerned, especially if he operates under a charter provision where the authority is limited to preventing, controlling, and extinguishing dangerous fires, and enforcing all laws relating thereto.

Even under the assumption that fires are involved, what action can a fire department take when it is physically obstructed from access to the fires, or is placed under attack by rioters while attempting to control such fires? A fire chief must be concerned about possible civil liability under such circumstances. As discussed earlier in this study there has been a new look by the California Supreme Court at the old doctrine of sovereign immunity. New statutes also have been added to the state's government code which have eliminated the traditional doctrine of immunity from tort liability which the fire fighting forces, like other "governmental" activities, have had applied to them for centuries past.

Definition of a Riot

It is important to know just what acts are considered sufficient to constitute a riot for several reasons. Examples are the effect they may have on avoiding insurance liability for riot damage, and the authority of the fire chief to take unusual measures to cope with rioters, as well as his possible liability for acting beyond the scope of his authority.

A riot has been defined as "disorderly behavior" in Vermont. The essential elements of riot are common intent and concert of action in the furtherance of such intent, according to a Georgia decision. But in Colorado the court said that concert of action is not essential, nor need there be shown a previous agreement of conspiracy. At common law there couldn't be a riot unless at least three people were taking part.

But in an Oklahoma case where two hooded men clad in the regalia of the Ku Klux Klan flogged and beat another person, the offense was held to be a riot, involving concert of action between three or more persons. In California a riot is defined in its Penal Code as: "Any use of force or violence, disturbing the public peace, or any threat to use such force or violence, if accompanied by immediate power of execution, by two or more persons acting together, and without authority of law, is a riot." Taking part in a riot is punishable by a sentence of up to one year in the county jail and up to $1,000 fine, or both.

The Penal Code (Section 197), in setting forth the conditions which make homicide justifiable, includes: ". . . When necessarily committed in attempting, by lawful ways and means, to apprehend any person for any felony committed, or in lawfully suppressing any riot, or in lawfully keeping and preserving the peace."

A Texas case says that unlawful assembly is a necessary element of an offense in engaging in a "riot." North Carolina agrees that unlawful assembly is required to constitute a riot, and adds two other ingredients; i.e., intent mutually to resist lawful authority, and acts of violence; whereas in South Carolina, the essence of the riot may be the *terrorem populi* (frightening the public) or it may be the committing of some unlawful act with violence, or in a violent or tumultuous manner.

A Connecticut court has declared that a riot, insurrection, and civil commotion can exist by virtue of temporary outbreaks of violence, without being classed as an organized rebellion against the government. A New York court held that a riot is more than a breach of the peace. It is a disturbance of the public peace, and implies the idea of a lawless mob accomplishing, or bent on accomplishing, some objective in a violent and turbulent manner that creates public alarm or consternation or terrifies, or is calculated to terrify, people.

A fire that was set secretly by disgruntled employees, without causing any disturbance did not constitute a riot within the meaning of the Oregon statute, so as to permit recovery under a "riot" and civil commotion insurance policy.

Where, however, a group of about thirty people in New York created disturbances, hurled missiles, and damaged property, notwithstanding the efforts of police, and about two hours later demolished a store front and pilfered merchandise, this was held to be a "riot" within the statute making the city liable for damage.

Where a Kentucky marshal, by an unauthorized act, burned down a hotel in order to arrest three persons for whom he had a warrant, the court held that the insurance company could not avoid paying the fire insurance claim on the ground that this constituted a "riot," which released it from liability under the policy's excepted risks of "riot," or, by order of any civil authority.

But in Indiana, where five masked men broke into a dwelling at night, forced the occupant to flee by threat of personal violence, and then burned the building, this was held to constitute a riot within the meaning of a fire insurance policy stipulating against loss or damage by fire caused by a riot.

There are also cases involving insurance company claims; e.g., an attempt was made to avoid liability under its burglary insurance because of the "riot" clause exception in Kirshenbaum v. Mass. Bonding & Ins. Co. (Nevada);[111] and where two men, with three other "lookouts," entered a clothing manufacturer's loft and in a "rowdy fashion" dumped acid on garments while the other intimidated employees with a revolver, this was held to be "riot attending strike" within the insurance policy.

In another New York case where at least two persons, without disturbing anyone, damaged a house by distributing creosote during the nighttime, this was held to be a "riot" within the meaning of the "riot" clause attached to the general insurance policy.

Fire Department's Participation in Riots

From a legal standpoint, the role of the fire department in civil disturbances is fire control—not riot control. This is not merely a matter of tradition or personal preference on the part of fire department administrators—it is a matter of law. As pointed out earlier, the authority of any city department stems from the city charter, and while it usually distinctly places the duty upon the police department to quell riots, it seldom even mentions riots in connection with the duties of the fire department; its primary job is concerned with fire prevention and control, and it would be an *ultra vires* act for its members to engage in any mob control activity, except of course as might be incidental to the suppression of dangerous fires, or in self defense.

Moreover, in California, as in some other states, there is a law which imposes liability upon the city for failure to discharge its duty (Government Code, Section 815.6), and therefore if a fire breaks out and the fire department is notified, the city could be held liable in the event that its fire fighters fail to exercise reasonable diligence to extinguish the fires, since it has the *mandatory* duty to do so.

It is hard to conceive, however, that any jury would find that fire fighters were not using reasonable diligence to fight the fire, merely because they were forced to withdraw from a conflagration area owing to physical attack (including gunfire) where there were insufficient police or soldiers to provide a measure of protection. It would certainly seem that the only practical thing to do when being attacked under these conditions is to pull back to a perimeter where one can protect exposures, and only attempt to gain access inside the riot zone where it is imperative to effect the rescue of persons who may be in danger of being trapped.

California's Government Code does have one section which might be applicable and somewhat helpful to fire fighters, in case they are ever sued for injuries caused by fire streams. This could happen since fire fighters have had to shut down streams on some occasions to help get people out of fire-involved

---

[111] 107 Neb. 368, 186 N.W. 325 (1922).

stores; had they not been made aware of their presence, very serious injuries could have resulted from the powerful ladderpipe and wagon battery streams. Section 850.4 of the Government Code provides that neither the city nor the fire fighters, *acting within the scope of their employment,* shall be liable for any injury caused by *fighting fires.* But this immunity would not apply to a willful or negligent failure to fight a fire, nor to injuries caused by fire fighters engaged in other duties, such as fire prevention, education, or training (which *are* within the scope of their employment); and it would certainly *not* apply to acts which are clearly *outside* the scope of their duty, such as mob control.

A final word of caution—the carrying of concealed weapons is a felony, and fire fighters should not attempt to carry out a policeman's duties unless required by law to do so. They will have more than enough to keep them fully occupied just fighting fires, without taking on the added responsibility (and possible liability) of fighting people.

Review Activities

1. Consider the case in which a deaf-mute was struck by a fire truck (206 Misc. 79, 132 Fink v. New York, N.Y.S.2d 172 [1954]). Then explain in writing how you feel fire truck warning devices might be designed so that blind, deaf, and mute (or any combination thereof) persons can react safely to the approach of emergency vehicles.
2. When fire apparatus and personnel respond to fires outside their own areas, who is held responsible should an act of negligence occur? Explain how a fire fighter is protected in the event of injury during a similar situation.
3. What constitutes a riot? Describe the role played by the fire department during a riot.
4. Today, who is held accountable for negligence on the part of a fire department? Why? Describe the kind of liability statutes your community's fire department has.
5. Explain how "ordinary negligence" and "wilful misconduct" apply to emergency vehicles. Cite an example of each.
6. Give examples of how alarm signal wires might be considered a source of liability to a city.

Chapter Seven
MUNICIPAL LIABILITY TO MEMBERS OF FIRE DEPARTMENT

At common law a municipality is usually held not liable for injuries to fire fighters, because fire fighting is a governmental function, but generally speaking a municipal fire fighter may recover damages for injuries caused by the negligence of the municipality if he is performing a ministerial function, e.g., flushing the streets, filling a swimming pool with water, repairing a cistern or a fire alarm system, or in performing any functions for which the municipality receives a profit rather than for public welfare.

Safety in the Fire Service
Safe Place to Work

There is a split of authority on the question of whether the city owes a duty to fire fighters to provide them with a safe place to work, but in those states applying the occupational safety standards to political subdivisions, under OSHA, discussed in Chapter Five, obligation to provide a safe place of employment will receive more attention.

In Bowden v. Kansas City[1] a fire fighter was injured when a horse tripped in a hole worn in the wooden blocks of the apparatus floor and stumbled upon him. In holding the city liable the court said:

"When a municipal corporation assumes the performance of a public duty which was permissive only and enters upon the discharge of such duty, and through the negligent performance thereof by one of its authorized agents one is injured either in person or property, the corporation cannot escape liability by saying that the performance of the duty was not imperative."

On the other hand, where a fire fighter broke his back when he fell from a faultily constructed drill tower, or where a fire fighter had to have his leg amputated as a result of injuries received when the weights on a fire station door broke, allowing it to fall and strike his knee, or when the driver of a fire

---
[1] 69 Ka. 587, 77 P. 573, at p. 594, 66 L.R.A. 181 (1904).

truck struck his head on a board nailed to the ceiling of the station when he was responding to a fire alarm, the courts held that the municipal corporation is not liable in the absence of an express statute so providing.

In a Texas case, a fire fighter (plaintiff) and another fire fighter were strapped together while the plaintiff leaned out to push up a pompier ladder to the next floor window of a drill tower. Due to some defect in the tower the plaintiff fell, taking the other fire fighter, who was standing inside the window, with him, with plaintiff receiving a broken back. No recovery of damages was allowed.

In Alabama, where a fire fighter was injured when he fell from an extension ladder, the court, in holding that the maintenance and operation of the fire department was a governmental function, said that the city is "exempt on reasons of sound public policy which would relieve them of the disadvantages and embarrassments of responsibility for those inevitable miscarriages which attend the performance of duties at once so difficult, so urgent, and so important. To hold otherwise might well frighten our municipal corporations from assuming the startling risk involved in the effort to protect themselves against fire." But as recent legislation and current opinion will indicate, the above argument is no justification for failing to provide an employee with a safe place to work and safe tools to work with. A fire fighter doesn't ask to be shielded from danger when he is fighting a fire, and is willing to assume all the risks incident to saving life and property that are attendant to the suppression of a conflagration, but he should be entitled to place reliance on the tools and apparatus with which the municipality furnishes him to carry out his duties. Thus the Kansas court said that regardless of whether the city makes a profit from the activity, the liability springs from the duty which is due every person, whether natural or artificial, to exercise such reasonable care in the conduct and management of his property that will not unnecessarily result in injury to another.

Safe Apparatus to Drive

There has been some attempt to impose municipal liability to fire fighters for injuries caused by defects in fire engines, as in LaFayette v. Allen[2] where a fireman was burned and crippled for life when an old worn-out steam engine exploded; other cases have held that in the absence of statute, neither a paid fire fighter nor a volunteer may recover from the city under such circumstances.

Thus, where defective brakes on a fire engine caused a fire fighter to be run over, and where a hose reel collapsed under another fire fighter, the city was not held liable for the consequent injuries. In the latter case, however, the dissenting opinion expressed the better view, though perhaps not the majority view, when it asserted that the city "should exercise reasonable care to furnish him with safe appliances for the performance of his arduous and dangerous duties," irrespective of whether he is a volunteer or paid fire fighter.

In states having a Public Liability Act, similar to the one in California, the municipality may be liable for injuries resulting from defects in fire apparatus,

---

[2] 81 Ind. 166 (1881).

since such statutes include liability for defects in personal property as well as realty, and have been construed to include defects in trucks.

Safe Equipment

In the case of MacClave v. New York (1965),[3] the New York Appellate Court upheld a State Supreme Court jury verdict of $145,000 damages to the widow of a fire lieutenant who died from carbon monoxide poisoning while wearing a filter type gas mask in a smoke-filled apartment.

Of special importance in this legal decision is that the City of New York was found negligent on two counts: First, because a few months before the lieutenant died in the fire, the fire department had issued the filter-canister mask "for general fire fighting purposes"; and secondly, because no means had been provided to enable the fire fighters to determine the carbon monoxide of the atmosphere. The Appellate Division held that the city was negligent in furnishing these masks to the fire fighters, making them available for use at fires, and in effect, encouraging their use when the limitations were apparent. The court declared: "The City in furnishing masks which could not be used under any circumstances, was guilty of negligence as a matter of law. . . ."

Another part of the court's comments has particular bearing on the use of breathing apparatus by fire departments in atmospheres for which the equipment was not designed. The court stated: "We must agree with the general proposition advanced by the defendant that when an appliance, furnished by an employer, is used for a task for which it is not intended, the employer is ordinarily not liable for injury resulting from such use, and, so, if this mask were used for purposes other than fire fighting, for example, for skin diving, that principle would apply, but concededly, it was issued by the city for general fire-fighting purposes, and it was intended by the city that firemen use it for that purpose, and it was in such use that it failed."

In California, the Attorney General has ruled that the Safety Orders of the Division of Industrial Safety (Department of Industrial Relations) apply to public employees, including those of a chartered city, and therefore fire fighters must be furnished turnout coats, helmets, boots, and other protective equipment needed for fire fighting.

Defects in Streets

A municipal corporation is liable for injuries received by fire fighters if they are caused by defects in streets, notwithstanding the fact that such injuries occur while in the performance of a governmental function. Thus, where a fire fighter was jolted off the hose wagon when the apparatus struck a hole in the street, the injury was found to be caused, not by the negligence of the driver in striking the hole, but in the city's negligence in not repairing it. In Kansas City v. McDonald,[4] where the driver of a hook and ladder wagon was killed when his truck hit an unguarded and unlighted pile of building materials left in

---

[3] MacClave v. New York, "Legal Decision on Filter-Canister Mask," *Firemen,* Vol. 33, No. 5, May 1966, p. 9.
[4] 60 Ka. 481, 57 P. 123, 45 L.R.A. 429 (1899).

the street with the permission of the city, the court declared that cities are required to keep and maintain their streets in a reasonably safe condition for public travel, and are held to as great a degree of care toward a fire fighter driving over same in the discharge of his duties as they are to any other traveler.

Statutory Liability to Fire Fighters

Though under the Municipal Corporation Law of New York, liability was imposed upon cities for injuries to fire fighters received in line of duty, no recovery could be had for the death of a fire fighter who was killed while riding in a fire truck, at the orders of his chief, that was on its way to obtain a grappling hook for recovering drowned bodies. The victim was held to have been acting beyond the scope of his authority. His authority only extended to the power of the fire department authorized by law, i.e., to extinguish fires and protect property from fire. Since his chief only had the power to order him to do intra vires acts, i.e., acts within the scope of his authority, the court said that he should have refused to do the act in question.

A state statute may impose liability upon a municipal corporation for injuries caused by the negligent operation of city vehicles, but the New York Court, in Ottman v. Village of Rockville Center,[5] held that such a statute does not apply to fire fighters killed in the course of duty, but is intended for the benefit of wayfarers and occupants of vehicles who are strangers to the immediate service then being rendered in the operation of the vehicle. The Ottman case held the fire fighter liable whose negligence was responsible, however, so that the widow of the deceased was not left entirely without a remedy. It would seem that in jurisdictions creating an immunity from civil liability for fire fighters when driving emergency vehicles that when an injury occurs under such circumstances, the statute should be construed to afford the injured fireman a remedy.

Fire fighters are protected in some jurisdictions by Workmen's Compensation Acts, which are dealt with in Chapter 9, and in some cities, under Employer's Liability Acts.

In construing the Oregon Employer's Liability Act to include a lineman attached to the fire department who was injured while clearing away poles with fire alarm wires attached, the court rejected the attempt to distinguish between employees engaged in a ministerial function, as was the case here, and those employed in a governmental function. The court declared that the mere fact that he takes an oath of office does not preclude him from the protection of the act which makes all corporations liable to a person engaged in a dangerous occupation:

"The employment was hazardous. Why it (the city) should not be liable in such a case, and yet liable if, through the negligence of the Street Superintendent, a citizen stepped into a hole improperly left in the sidewalk, passes understanding. But the present is a stronger case because here we have the Employer's Liability Act to reinforce the common law liability. If a policeman,

---
[5] 273 N.Y. 205, 7 N.E.2d 102 (1937).

hurrying to overtake a criminal, had fallen into a hole in a street, negligently left unlighted by the street authorities, it would be considered a poor defense to say, 'Oh he was a public officer.' The attempted discrimination between a quasi-public officer and an employee is purely technical, and the city cannot escape liability by requiring an employee to take an oath and then calling him an 'Officer.'"[6]

Review Activities

1. Explain why, in those states having Public Liability Acts, the municipality may be liable for personal injuries resulting from defects in fire apparatus.
2. (a) If a fire fighter were using a city fire department ladder to paint his own home and the ladder collapsed from old age, would the city be considered responsible for injuries suffered by the fire fighter? Why or why not?
    (b) If this same situation occurred at the scene of a fire, would the result be the same? Why or why not?
3. On what occasions can a municipal fire fighter recover damages from the city?
4. Based on what you have learned from this chapter, write a general statement that explains your feelings concerning municipal liability to members of fire departments. Include in your statement your feelings concerning the provision of a safe place to work. (Consider that while on the job, much of a fire fighter's time is not spent at the scene of a fire.)
    (a) When you have completed your statement, exchange it with the statements prepared by three or four of your classmates.
    (b) Discuss with these same classmates any differences in opinion, and how such differences might be resolved.
    (c) In an open class discussion, consider how municipal liability to members of fire departments might be standardized on a nationwide basis.

---

[6] Asher v. Portland, 133 Ore. 41, 284 P. 586 (1930).

Chapter Eight
FIRE PREVENTION BUREAUS

Source of Power

A function of increasing importance among fire department activities is the work of preventing fire. To carry out the duties imposed by charters of large cities, fire prevention and public safety bureaus have been created. As in other fields of safety, the "Three E's" (Enforcement, Engineering, and Education) are also a means of advancing firesafety.

Since "enforcement" requires power and effective laws to enforce, to find the powers which a fire prevention bureau may exercise, one generally looks to those given the fire department by charter or statute, and to the specific powers given the bureau by ordinance. For example, by express provision of the charter, in Los Angeles the fire department "shall have the power and duty to control and extinguish injurious or dangerous fires and to remove that which is liable to cause such fires, and to enforce all ordinances and laws relating to the prevention or spreading of fires, and all ordinances and laws pertaining to fire control and fire hazards within the city of Los Angeles. . . ." In addition, municipal ordinances of Los Angeles authorize the chief of the fire department to make and enforce such rules and regulations for the purpose of prevention and control of fires and fire or explosion hazards, as are necessary to carry out the purposes and intent of the fire ordinances.

The power granted by the Los Angeles Charter extends to the enforcement of ALL laws and ordinances relating to the prevention of fire. Hence, whether regulations may be found in sections of the municipal code dealing with building requirements, or the sections relating to health requirements, or even in the laws of the state, the fire department has the charter-given power to enforce them if they relate to the preventing or spreading of fires.

Thus, in Los Angeles and most large cities, the fire department or its fire prevention bureau has the power to do everything necessary to make the city safe from fire, for even without such enabling clauses in the charter or in ordinances, such power would exist as a reasonable exercise of the police power, the latter being inherent in every state and its legal subdivisions. The late Robert S. Moulton, when he was technical secretary of the National Fire Protection Association, made this statement:

"Following the Cocoanut Grove fire, the officials concerned were quoted in the newspapers as stating that the existing codes and laws were not adequate to assure safety in places of public assembly such as this night club. We cannot believe, however, that this disaster (in which almost 500 lives were lost) is chargeable to any deficiency in the law.... In our opinion building and fire officials can now do practically everything that is necessary to assure public safety from fire without any more laws.... Even though there may be no further authority than a city charter, these officials can issue orders for the purpose of safeguarding the public and in all probability have them upheld by the courts."[1]

In discussing the broad authority often delegated to a fire marshal in state laws and local ordinances, Judge Virgil Langtry, in the October issue of the NFPA Quarterly for 1963, had this to say:

"I believe that a general delegation of authority to a fire administrator can be upheld, even though a finger cannot be pointed at a grant of power to do each specific thing, so long as the official does not discriminate between people or corporations in the same class, does not set up discriminatory classes which are not based on reason, that his action is reasonable as distinguished from being arbitrary, and his actions are directed toward achieving the desirable intentions of a fire code.

"If the fire marshal can confine his action within these boundaries, and he has an attorney who will represent the trend of liberalizing decisions with the logic which undergirds them, he may expect to take valuable action in the field of fire prevention, even though he does not find it specifically mentioned in his fire or building code."[2]

A concise discussion of the practical limitations of fire prevention and building codes, insofar as the necessity of code enforcement officials to decide what constitutes a safe building, is contained in a *Fire Journal* article, "Discretionary Powers in Code Enforcement."[3] The author reviewed a number of guidelines and documents available to help the officials make sound decisions.

Fire Prevention Laws

From time immemorial man has recognized the fact that uncontrolled fire is a menace to society. Hence certain restrictions have necessarily been imposed on the conditions of its creation and use. At first, laws in the field of firesafety were very general in their nature, for it was thought to be common knowledge that careless conduct with respect to fire leads to disastrous results. Soon, however, it was found insufficient to merely lay down the rule that "no person shall create a fire hazard," since it was found unreasonable to expect anyone to know, under all circumstances, every condition that creates a risk from fire.

Although the legal requirements differ in various localities, there has been a marked tendency to enact legislation in the form of federal and state laws,

---

[1] Moulton, Robert S., "Fire Inspection," NFPA *Quarterly*, Vol. 36, No. 3, Jan. 1943, pp. 197–198.
[2] Langtry, Virgil, "The Fire Marshal's Delegated Authority," NFPA *Quarterly*, Vol. 57, No. 2, Oct. 1963, pp. 126–129.
[3] Bond, Horatio, "Discretionary Powers in Code Enforcement," *Fire Journal*, Vol. 60, No. 1, Jan. 1966, pp. 40–41.

municipal ordinances, and rules and regulations having the effect of law to overcome the effect that hazardous materials and poor building construction have on life and property.

Among the federal laws dealing with firesafety, in addition to legislation for marine protection from fire as required by U.S. Coast Guard regulations, and the Department of Transportation's regulations for interstate carriers with respect to explosive or other hazardous materials, will be found the laws dealing with forest fire control and flammable fabrics. In recent years, federal regulations for firesafety has been greatly increased.

Although state laws vary considerably, those which are most commonly found relating to fire departments are arson laws, fire marshal acts, criminal fire regulations, as well as civil fire regulations. In most states, the detailed legal requirements of the state as to the control, prevention, and spread of fire, are generally formulated as rules and regulations by a state fire marshal. Other state laws relating to fire protection may authorize the formation of fire departments in cities, towns, and other local districts, or relate to forest fire control, as well as to fire fighters and fire departments.

In addition to the federal, state, and county laws, there are municipal ordinances affecting firesafety. They are usually found as electrical codes, building codes, and fire prevention codes.

Scope of Police Power

Laws governing the conditions that give rise to dangerous fires are subject only to the ordinary limitations on the exercise of the police power. Chief Justice Shaw in 1853 gave one of the earliest definitions of the police power, when in Commonwealth v. Alger he said:

"We think it is a settled principle, growing out of the nature of well ordered civil society that every holder of property, however absolute and unqualified may be his title, holds it under the implied liability that his use of it may be so regulated, that it shall not be injurious to the equal enjoyment of others having an equal right to the enjoyment of their property, nor injurious to the rights of the community."[4]

The court points out "that this is very different from the right of eminent domain—the right of a government to take and appropriate private property to public use whenever the public exigency requires it; which can be done only on condition of providing a reasonable compensation therefor." The requirement of compensation is a constitutional provision (14th Amend.), though the right is said to be implied as a necessary incident to the expressly granted powers.

The broad scope of the police power of the state was asserted by Chief Justice Redfield's opinion in a case, when he said: "This police power of the state extends to the protection of the lives, limbs, health, comfort and quiet of all persons and the protection of all property within the state."[5] This was amplified even more by the United States Supreme Court when it stated: "We

---
[4] 7 Cush. 53 at p. 84 (1853).
[5] Thorpe v. Rutland Ry Co., 27 Vt. 140 (1855).

hold that the police power of a state embraces regulations designed to promote the public convenience and general prosperity, as well as regulations designed to promote the public health, the public morals, or the public safety."

The police power has been frequently used to promote firesafety in various fields of legislation. Better fire protection has been one of the major grounds in upholding the constitutionality of zoning ordinances, set-back ordinances, height limitations for buildings, the bulk and area of buildings, the location of oil wells and garages, and the prohibition of billboards.

A municipality has, therefore, ample authority through the exercise of its police power, as well as through the rights granted to it by its charter, to deal with arson, incendiarism, nuisances, building construction, location of hazardous industries, occupancy, and numerous other important factors relative to the prevention of fire.

A municipality must not act arbitrarily or unreasonably in administering valid ordinances; however, the laws and regulations themselves should be nondiscriminatory and bear a reasonable relation to the objectives sought to be obtained.

In 1965, ordinances restricting to 1,500 gallons the capacity of tank vehicles making deliveries of gasoline to service stations were held unconstitutional in Missouri and Wisconsin. In the Wisconsin case the court said that "the transportation of gasoline in quantities from six to eight thousand gallons provides a greater degree of safety to Tomah than the provisions of this ordinance."[6]

The Supreme Court of South Carolina held that an ordinance restricting gasoline tank trucks to 1,250 gallon capacity was unconstitutional.[7] A Texas decision declared an ordinance unconstitutional which, while it didn't limit the size of the tank truck, limited its load to 1,500 gallons.[8] A Colorado case, which arrived at a similar decision, was cited in the preceding case. The Wisconsin Supreme Court summed up the reasoning in one of these ordinance cases as follows:

"Local ordinances should have some reasonable resemblance to recognized national standards established by qualified organizations, or otherwise the regulated industry would be at the mercy of every whim and caprice of the many different communities. . . . It is not for the Court to substitute its notions as to what is fair and reasonable, but to ascertain under all existing circumstances whether the standard established is, in fact, reasonable and bears a reasonable relation to the public health, welfare and safety of the city and its inhabitants."[9]

There are several decisions bearing upon the scope of the police power and its limitations as applied to fire prevention. Ordinances fixing the boundaries of fire districts have been met with little objection, and early cases have upheld the power to destroy buildings to prevent the spread of fire.

In New York the court asserted that "it has been repeatedly held that the fire commissioner has authority under the ordinances to make an order re-

---
[6] Clark Oil and Refining Corp. v. City of Tomah, 30 Wis.2d 547, 141 N.W.2d 299 (1966).
[7] McCoy v. Town of York, South Carolina, 8 S.E.2d 905 (1940).
[8] *Ex Parte* O. R. Rodgers, Tex. Cr. App., 371 S.W.2nd 570 (1963).
[9] City of Colorado Springs v. Grueskin 422 P.2d 384 (Colo. 1966, Rhg. den. 1967).

quiring the owner of a building used for manufacturing purposes to install an automatic sprinkler system to extinguish fires."[10]

A California statute giving the owner of property adjacent to frame buildings within a fire district the right to bring a suit for any special damage suffered as a result of their maintenance has been upheld, but a Nebraska law, making the finding of the fire marshal conclusive upon the owner that his building was a nuisance, was held unconstitutional as denying a notice and hearing under the "due process" clause of the Fourteenth Amendment to the Constitution. The Nebraska court said that where a building is a public nuisance because of age or dilapidation, making it liable to fire, such steps may be taken as are reasonable and necessary to abate the nuisance. But the decision of the fire marshal's department that the building is a nuisance, without a judicial determination of some kind on the question, is not conclusive on the owner or final upon the "courts."[11] Similar statutes have, however, been sustained in other jurisdictions. One view sustains the power to take action to declare and abate a nuisance without the necessity of a formal hearing, as a reasonable exercise of the police power, but liability may arise for destruction of property which is later found not to have been a nuisance in fact. The best procedure requires a notice and hearing in all cases where emergency action is not necessary.

In a California case the plaintiff was unsuccessful in his attempt to enjoin the demolition of a frame building even though the city had recently given the plaintiff a permit to have a large expensive oven installed in the building.[12] The United States Supreme Court found that the building had been unlawfully erected within the fire district, and that its demolition was not an unreasonable exercise of the police power; further, that the city could not be estopped by any inconsistent conduct because the police power cannot be contracted or bartered away; nor was there impairment of any obligation of contracts that the owner might have had with the tenants, for all contracts are entered into with a reasonable exercise of the police power as an implied condition.

In a Montana case, the facts were, briefly: the defendant owned a forty year old, two-story, frame dwelling house in a residential district that was in general disrepair. The state fire marshal was empowered to inspect and, in proper cases, to condemn any building which "for want of proper repair, by reason of age, dilapidated condition, . . . or for any other cause or reason is especially liable to fire," and which "is so situated as to endanger other buildings and property," as a public nuisance and to order the condition remedied; but if the order is not obeyed, the fire marshal may not summarily proceed, but can only maintain an action against the owner "for the purpose of procuring an order from the court," along the same lines as the order given the owner by him. The building in question, in addition to having been partly damaged by fire, was an "eyesore" to the neighborhood, although it was not located closer than forty feet to the nearest adjacent building. A twenty day notice was given and ignored, so a complaint was filed setting forth the facts about the building,

---
[10] People v. Miller, 165 N.Y.S., 176 N.Y.S. 206 (1917).
[11] State v. Keller, 108 Neb. 742, 189 N.W. 374 (1922).
[12] Maguire v. Reardon, 41 Colo. App. 596, 183 P. 303, 255 U.S. 271 (1921).

and asking that it be declared a public nuisance and either repaired to remedy its dangerous condition, or torn down and removed.

The Montana Supreme Court held the measures constitutional, but warned that "such a statute should be administered with caution and for the purpose of fire protection alone, and not to promote the 'city beautiful' idea." The court said, "we cannot overthrow a judgment based upon a preponderance of substantial evidence following the wording of the statute, even though we may suspect that the real purpose of the proceeding is to relieve the neighbors of a local 'eyesore,' or be inclined to believe with the witness for the defense, that the building is no greater fire hazard than it would be if repaired."[13]

A Chicago ordinance requiring that "fire drills shall be conducted according to rules, methods and regulations prescribed by the Chief of Fire Prevention and Public Safety in buildings and portions of buildings of the following classes . . ." was held to be an unreasonable exercise of power granted the city in a fire prevention statute regulating industrial and manufacturing establishments. The court held the statute only authorized the preventing of fires before they start, and not to deal with the situation after the fire has started, as the fire drill ordinance attempted to do. A reasonable fire drill ordinance, where for example, its application would extend only to schools and occupancies having large public assemblies, would no doubt be sustained as a valid exercise of the police power. Note that the above decision did not consider the question of the validity of a school fire drill ordinance, but merely whether the particular statute authorized an ordinance requiring manufacturing and industrial occupancies to form fire brigades and hold fire drills once a month.

Under a statute giving the city of Mobile full police powers within the limits of the city, an ordinance making it the duty of the chief of the fire department to assign a fire fighter to all performances in any theatre, and making it compulsory for the manager of the theatre to pay the salary of such fire fighter was held a reasonable exercise of the police power. The court said that if the defendant suffers injury, it is *"damnum absque injuria."*

However, the Illinois court arrived at a different conclusion when confronted with the constitutionality of such an ordinance. The court said that a theatre is not a nuisance per se, and a declaration by a city would not make a nuisance unless it was such in fact. "The charge is not an inspection fee, which may be lawfully imposed upon the principle that it is compensation for services rendered beneficial to the person required to pay it," the court said.

Elimination of Nuisances Constituting "Fire Hazards"

Some conception of the broad powers that may be exercised in an effort to enact and enforce firesafety regulations can be gathered from the foregoing cases. There is another field, however, which may prove a source for effective

---
[13] State *ex rel* Brook v. Cook, 84 Mont. 478, 276 P. 958 (1929).

fire prevention work, and that lies in the elimination of nuisances. A fire hazard is but another name for a public nuisance, and the methods of removing them are the same. It is now well settled that muncipalities may enact ordinances declaring that certain conditions shall be deemed to create a nuisance, and under its police power provide for its abatement. It has been held, for example, "that the state may order the destruction of a house falling to decay or otherwise endangering the lives of passers-by; the demolition of such as are in the path of a conflagration; the prohibition of wooden buildings in cities; the restriction of objectionable trades to certain localities."[14]

In Kaufman v. Stein[15] the court stated that a "wooden structure is not a *nuisance per se*. It is the circumstances that make it a nuisance. Even when they are originally built in a place remote from the habitations of men, or from public places, if they become nuisances by reason of roads afterwards being laid out in that vicinity, or by dwellings subsequently erected within the sphere of their effects, the fact of their existence prior to the laying out of the roads or the erection of the buildings is no defense."

Care must be taken to allege sufficient facts so as to indicate to the court that a particular building does in fact constitute a nuisance, for "if such a building is a nuisance at all, it is a *nuisance per accidens*—because of its use, location, surroundings or other circumstances." Where several witnesses testified that "by reason of its dilapidated condition, loosened siding, curled shingles, decaying beams, being without a substantial foundation, with openings in the roof and debris inside, this building was especially liable to fire from outside causes, spontaneous combustion from damp and decay, vandals entering and lighting matches, mice and rats coming in contact with matches left in the building and the like, and that if it did burn, by reason of its condition, sparks, burning shingles, and materials were likely to be carried much farther than the nearest buildings . . . ," and where expert evidence was also adduced to the effect that the building was so far gone to ruin and decay that it could not be repaired, the court found these allegations sufficient to describe a public nuisance that could be abated by the fire marshal.

Where the lower court in California, without expressly declaring the building to be a public nuisance, ordered it abated in an action instituted by the San Francisco City Attorney against the landowner, the Appellate Court said, "The trial court did in terms find, however, that said stucture was old, dilapidated, unsafe, defective, abandoned, and extremely dangerous; that the chimney thereon was dangerous, extremely defective and in an unsafe condition; that practically the entire structure was rotten, broken, did not conform to building laws, and was a menace to life and limb of passers-by and to the property adjacent thereto; furthermore that it was an extreme fire hazard and unsanitary. Clearly the above findings warranted the trial court in directing that the structure be abolished."

In addition to actions that may be brought by public authority, it is well to point out to the property owner whose buildings constitute a nuisance, the

---
[14] Lawton v. Steele, 145 U.S. 133 (1894).
[15] 138 Ind. 49, 37 N.E. 333 (1893).

possibility of civil liability to private individuals for the maintenance of a nuisance. In a Pennsylvania case it was said: "It is true that a private person not specially aggrieved cannot abate a public nuisance, and especially where a statute provides a remedy for an offense created by it, that must be followed. It is well settled, however, that a private person, if specially aggrieved by a public nuisance may abate it. . . . If the owner or tenant of a powder magazine should madly or wickedly insist upon smoking a cigar on the premises, can anyone doubt that a policeman or even a neighbor could justify in trespass for forcibly ejecting him and his cigar from his own premises?"

A California court said that a six-story, nonfireproof building creates some fire hazard to the community and adjoining property, but where its erection is not illegal, no right of action on the ground of the increased hazard arises; there must be shown some exceptional damage. This case affirms an older case where after showing that three fires had occurred in an adjacent pottery factory since 1917, plaintiff succeeded in 1932 in obtaining an injunction against operating a portion of the plant on the ground that it presented a constant fire hazard to his residence.

Statutes in many states grant anyone the right to abate a nuisance, where it can be done without a breach of the peace.

Professor Willoughby, in one of his works on constitutional law, has the following to add on the subject of nuisances: "The aid of equity may be invoked for the abatement of nuisances whether public or private, or, in private nuisances, the individual may, at his own risk, himself undertake the abatement. In either case, whether by a process of the court or by personal action, notice to the person responsible for the nuisance is not required where there is an emergency. This constitutes an exception to the general principle that private property may not be taken or destroyed except after judicial proceedings and condemnation with notice to the owner."

As far back as 1927 it was confidently asserted by the authors of a scholarly treatise on fire prevention bureaus that the municipal exercise of the power of demolition of dilapidated buildings and the power to demand improvements in existing buildings was everywhere being recognized, and there is cited a great number of cases in various cities where such actions were upheld.

Retroactive Measures

How far may a fire prevention bureau go in asking for the installation of safety features, the requirement of which has been enacted subsequent to the erection of the building? Some legislation has been expressly made retroactive, and its validity upheld. For example, it is commonly required that stairways and vertical shafts in buildings over two stories in height used for factories and living purposes be enclosed, and that new fire escapes and fire doors be installed on existing buildings even though the latter were built in accordance with the laws then in existence. It has been held, however, that an ordinance cannot be made retroactive with respect to the removal of existing buildings from the fire limits, although such structures may be barred from being moved into the fire limits.

Another significant step is the widespread adoption of laws requiring all wood shingle roofs be replaced within a given period. The retroactivity concept is merely based on the premise that property should not be taken without "due process of law." If notice and opportunity to be heard are provided, reasonable police power measures which increase the public safety and reduce the danger from fire are valid even though they may require changes in existing structures. "Retroactive legislation which does not impair vested rights or violate express constitutional prohibitions, is valid. . . ." An example of a case upholding retroactivity of a building code is a District of Columbia decision that meant that about 23,000 buildings would have to comply with requirements for fire walls, fire doors, alarm systems and other precautions.[16]

Ralph P. Dumont, an attorney in New London, Conn., summarized the problem of retroactivity as follows: "Actually a law is not unconstitutional, as a general rule, merely because it is retroactive in application or operation. There are only four basic situations in which retroactivity might be a proper reason for condemning a law:

1. An *ex post facto* law;

2. A law impairing the obligation of contract, but this refers only to state laws;

3. A law violating the Fifth or Fourteenth Amendment, because it takes property without due process of law; and

4. A law of a state violating the Fourteenth Amendment's requirement of equal protection of the law.

"*Ex post facto* refers only to criminal law . . . insofar as the second point is concerned, it's hard to see how an existing building is a contract. We are left, therefore, with the problem of due process and of equal protection of the laws.

"'If the regulation appears to be reasonably calculated to correct the evil it was aimed at, and if it does not impose too great a burden on private rights, it is clear that the courts will uphold the law and, in effect, sanction its application retroactively.' This is exactly what the United States Supreme Court did decide in the Queensides Hills Case. Queensides Hills Co. v. SAXL (328 U.S. 80, 1945).

"In that case the court dealt with amendment to the New York State multiple dwelling law which required automatic sprinklers in certain types of lodging houses built before the date on which the law was enacted. In fact, the law had no application to any but existing lodging houses. Mr. Justice Douglas wrote the Court's opinion and concluded that the statute was constitutional. He indicated that the public interest in protection from loss of life by lodging house fires was so great that it outweighed any private or vested right of the owners of the lodging houses. His opinion considered that as much as $7,500 had to be spent on properties worth only $25,000, in order to comply with the law.

"Furthermore, the opinion went on to dispose of the question of equal protection of the laws. The court ruled that it could not look behind the legislative determination of what was the best way to protect human life and property, nor

---

[16] Jones *et. al.* v. District of Columbia (Cir. 1963).

would it strike down the law merely because it was not applicable to all lodging house owners. Mr. Justice Douglas indicated that the legislature could make a distinction in treatment of old and new buildings and that it could make even stricter requirements for old buildings than new ones, if it wished so to do in its wisdom.

"Now, I think you'll find that courts in New England are saying about the same thing. Thus, in Windsor v. Whitney, 95 Conn. 357 (1920), Connecticut Supreme Court of Errors said, in effect, that the State may regulate buildings and lots retroactively in the interest of health or fire safety and for other related purposes. The free exercise of property rights, therefore, cannot be permitted and will not be permitted if detrimental to the public interest. Keep in mind that public safety can even limit the exercise of such great rights as free speech— one has not right to cry 'fire' in a crowded theatre.'

"The Massachusetts Supreme Judicial Court said, 'All contract and property rights are held subject to the *free exercise* of the police power.' Paquette v. Fall River (1958). It is interesting to note that the Queensides Hills case is cited and followed by the Supreme Judicial Court. Furthermore, we find the New Hampshire Supreme Court seemingly accepting the law of the Queensides Hills case also.

"There seems to be little question that most fire safety laws are, therefore, valid exercises of the police power and that they apply retroactively as well as prospectively. Nevertheless, each statute must be read carefully and you must look to your State constitutions as well.

"I, therefore, join with Judge John M. Wise of Michigan who . . . urged the evenhanded enforcement of the law against all violators. He said, 'Enforcement then is the key—good, hard-hitting, reasonable . . . enforcement.'"[17]

From a practical standpoint, the best method of correcting fire hazards that involve the expenditure of large sums of money is to put aside the question of whether the measures can be validly required under existing laws and to "sell" the person on the importance of installing the recommended safeguards to protect himself and his investment; the legal technicality of retroactivity is less apt to be brought into question if the law itself is not quoted to the person who is being "sold" on the measures. By stressing the underlying principles of fire protection, such as the subdivision of large areas, the reduction of high values in one area, the protection of vertical openings, etc., and by convincing the person of the benefits which will accrue from complying with the recommendations, e.g., increased life safety to employees, reduced insurance premiums, and lessened risk of civil liability, the installation of fire protective measures can often be obtained which otherwise may be legally impossible.

Many are not aware of the fact that their inusrance may be voided by acts which increase the hazard, even if these acts were accidental. For example, the insurance company did not have to pay for the loss of the plaintiff's goods in a storage building fire because the policies provided for nonliability for loss occurring "while the hazard is increased by any means within the control or

---

[17] Dumont, Ralph P. "The Doctrine of Retroactivity and Fire Laws," *Firemen*, Vol. 30, No. 10, Oct. 1963, pp. 32–33.

knowledge of the insured." About a month before the fire, the building which the insured rented for storage of the goods collapsed to such extent as to leave spaces between the walls, rendering it susceptible to increased drafts and to entry by children and vagrants.

The opinion which, applying the law of New Jersey, and ruling that the evidence was sufficient to justify a finding by the jury that the fire risk was increased by the partial collapse of the building, held that the fact that the hazard was increased by accidental means, and not by act or instrumentality of the insured, did not preclude suspension of the policy under the "increase of hazard" clause.[18]

The policy of the Los Angeles Fire Prevention Bureau, of making recommendations without giving undue emphasis to the question of retroactivity, is evidenced in one of its publications wherein a handbook for the guidance of school administrators and custodians is worded in a normative manner, directing what should be done to achieve the maximum of fire safety, without reference to any ordinance or statute.

Citizens' Personal Liability

Fire departments have found it helpful to constantly stress the personal civil liability that may arise as the result of the public's failure to heed their recommendations. That this is no mythical liability is amply illustrated by the numerous instances where large sums of money have had to be paid in claims following disastrous fires. Damage suits of seventeen fire fighters and survivors were settled by a paint company in Pennsylvania, amounting to over $116,000. In 1950 a circus company paid off the last of a $4,000,000 set of claims by the victims of its blazing tent in 1944. Damage claims numbering about 157 and totalling over $3,000,000 were settled by a large hotel where 119 persons died in a fire in the mid 40s. Suits for more than $2,000,000 were filed by the representatives of the 75 victims who perished in the Effingham, Ill. Hospital fire of April 5, 1949.

Roger Arnebergh, former City Attorney of Los Angeles, has commented thus on legal liability:

"Insurance against legal liability for the spread of fire is not in general use. To date, neither the fire nor the casualty companies have made any considerable effort to publicize and create a market for this protection and the result is that the public is largely ignorant of the existence of the hazard. If the public were fully aware of the hazard, the services of the fire inspector would be welcomed."

There seems to be an increasing trend on the part of insurance companies after paying loss claims to the insured party, to search for some possible defendant whom it can sue to recover the amount paid out. For example, after paying the insurance claim of the firm which had suffered a loss of its rayon yarn in a warehouse fire, the insurance company was awarded a judgment against the owner of the warehouse who was also the operator of a cotton gin

---

[18] Foldman v. Piedmont Fire Insurance Co., 198 F.2d 712 (Cir. 3rd 1952).

about 50 feet south of the warehouse and separated from it by a lot. Because considerable quantities of waste material from the gin had piled up near the door of the warehouse, the court found the defendant negligent, and as a bailee for hire, liable for want of ordinary care.[19]

The sources of civil liability with respect to fire originate both in the common law and in statutes.

Common law liability exists for injuries resulting from dangerous conditions of which the owner or occupant of the premises has notice. The general rule is stated in *Corpus Juris* as follows: "Notwithstanding a fire may be accidental in the sense that it is not intentionally kindled by him, one may be liable for injury occasioned by it where it is due to his negligence, or he has failed to use ordinary care and skill to extinguish it, or to provide adequate means for so doing."[20]

The important question then becomes, what is the measure of care that is required? With respect to the occupant of property the duty of care varies with the relationship the occupier bears toward the injured party. The duty owed business invitees is greater than that owed mere licensees or trespassers. For example, suppose a fire breaks out in a hotel. There would be a high degree of care owed by the management to see that a paying guest is promptly notified in time to make any possible escape, that the route of egress has been posted in the room, that the halls have been provided with sufficient light to find the means of egress, and that access to the outside is unobstructed by locked doors, furniture, etc. To the individual who is not a paying guest of the hotel, but merely permitted to sleep in a chair in the lobby as a humanitarian gesture, only the duty owed a licensee would be required—a duty that does not extend to taking elaborate steps to protect the individual but merely warning him of any hidden dangers and to not actively increase any hazards already present. In this example, such a duty would undoubtedly extend to notifying the person of the fact that there was a fire and of maintaining any safeguards ordinarily required by law, but would not obligate the management to see that he got out quickly or to underwrite the safety of his belongings. In the case of a trespasser, the duty would extend to notification of the danger only if it was known that he was there.

Civil liability can be imposed for the maintenance of fire hazards that are dangerous conditions on the grounds that they are actionable nuisances. This subject was discussed in previous pages, but it is well to point out that activities held to be a nuisance are always unlawful and subject to abatement either through legal process or sometimes by self help.

An NFPA *Quarterly* article discusses various lawsuits which involved death and injuries in fires. The case in which two women recovered over $150,000 damages against the owner of a Los Angeles dice manufacturing building having inadequate exits is presented in great detail.[21]

---

[19] Westchester Fire Insurance Company v. Atmore Truckers Association, U.S. District Court, 120 Fed. Supp. 7 (1954).
[20] 45 *Corpus Juris* 852.
[21] Bagot, M. H., "Civil Recourse in Fire Losses," NFPA *Quarterly*, Vol. 51, No. 3, Jan. 1958, pp. 206–212.

The liability of manufacturers, wholesalers, or anyone in the commercial distribution chain, to persons injured on account of the flammability of a garment is discussed in a March 1967 *Fire Journal* article. After pointing out that judgments in such cases can be great, owing to the enormity of the pain which the victim suffers (as high as $5,000 per day), the author cites a case where one pajama manufacturer was ordered by a jury to pay a housewife $140,000 for burns sustained when her pajamas brushed against a kitchen burner and caught fire. In these cases, compliance with the *Federal Flammable Fabrics Act* does not automatically mean that the industry is exonerated from liability to a person who is injured as a result of a garment's burning. The author said:

"The garment, even though it complies with the *Flammable Fabrics Act*, may have characteristics rendering it inappropriate for use in a child's dress or suit, or in something designed to be used in the area of the kitchen stove, or by the aged and infirm. The jury decides the facts in lawsuits—and they can decide that the manufacturer did not exercise prudence in using a particular cloth for a given garment even though that cloth was manufactured to pass the requirements of the *Flammable Fabrics Act*."[22]

Notice of a Hazardous Condition

Suppose a person were sprayed with atomized oil under high pressure issuing from a broken hydraulic line connected to a plastic moulding machine, and suppose that, as so often happens, the spray breaks into flame and burns the person. Would the owner of the machine be liable to the injured person for the damages suffered? A person not trained in the law would likely give a snap judgment without getting all the necessary facts. First it should be learned what the relationship of the injured person was to the management of the company to find out what duty, if any, was owed the injured party. Secondly, it is necessary to know whether the person in control of the machine was in any way negligent, i.e., did he have any idea that the line was weak and about to break and then fail to do what an ordinary prudent man would do—replace it. If he could have learned that it was defective by routine inspection but had never made inspections with the frequency ordinarily carried out by the others in that line of work, then it could be considered negligence even if he did not actually know that the line was defective, for he should have known it. When a person by using ordinary care could find out something if he took the trouble, he is said to have *constructive notice* of the condition. If a fire inspector had been through the plant and had warned the management that the use of that particular type of machine was unsafe unless a nonflammable oil were substituted for the flammable type, and after describing a few of the fires and burns recently caused by the failure of such high pressure lines, had given them a written recommendation to either use rigid piping with all-metal swivel couplings to eliminate flexible tubing, or else use one of the approved nonflammable hydraulic or heat transfer fluids, then the man-

---

[22] Elkind, Arnold B., "The Clothing Fire Problem: Some Legal Aspects," *Fire Journal*, March 1907, pp. 11–14.

agement would have had definite notice of the hazard, irrespective of whether or not it made its own regular inspection. The fact that the law may not have actually required the changes recommended by the inspector is not material to the issue of whether or not there was notice of the hazard. If after such notice, whether actual or constructive, a reasonable time had elapsed in which to correct the condition, but no action had been taken, then any jury might well find that the management had been negligent.

As pointed out previously, the first inquiry would be to see whether or not the person injured was in a class of persons to whom the management owed a duty of care, and to see just what degree of care would be expected under the circumstances. If the person were an employee, there is a duty to provide reasonably safe equipment and premises. Of course, if the employee had seen the lines break on previous occasions and knew how dangerous the work was, then there would be presumed to have been an *assumption of risk* in many jurisdictions, but he would be entitled to recover under the typical workmen's compensation statute now in effect in most states.

If the person injured were a salesman, customer, or other person who might be considered to be there on a matter of mutual interest to the company, then he would be classed as *business visitor* or *invitee*, and would clearly be entitled to protection against such negligence provided that it was customary for him to go into the shop area to transact his business. If he were not allowed to go into the shop and had been so advised of the restriction and the possible danger, then he would be in the position of either a *licensee* or a *trespasser*, depending upon whether the management condoned his presence in the shop on the same basis as might be done with a man who cuts through their premises on the way to his own plant, i.e., a licensee, or whether the management would actually eject him from the shop area if he had been seen there and would have no more tolerated his presence in that part of the premises than that of a trespasser.

While the duty to make inspections to discover defective machinery that might give rise to injuries extends to invitees, it does not extend to licensees or trespassers. If the licensee who cuts through the premises for his own convenience is warned of moving cranes or any dangers which are not obvious, and ordinary care is exercised, that is sufficient; he takes the premises as he finds them. And if he were the victim of the sprayed flaming oil it is extremely doubtful that a jury would permit him to recover for his injuries unless the fact that the line was about to burst had long been brought to the attention of management who had neither done anything about it nor warned him of it.

The *trespasser* is in an even poorer position to claim that a duty of care toward him had been violated. The duty extends only to warning of hidden traps and the like. If a man were blasting on his property, and he saw a trespasser cutting across his land in an area where he might get hurt, he would be required to warn him of the danger. Where a person stays in a zone of danger voluntarily (and not as a part of his duty), any injury suffered as a result would be considered due to his own negligence; the latter would be called *contributory negligence* if the other party was also negligent, and contributory negligence is a bar to recovery of damages in most situations.

Absolute Liability

Thus, it can be seen that the above principles are more complex than appears at first glance. For further example, suppose a person in the fumigation business were to store cylinders of methyl bromide or hydrocyanic acid on his premises. These are known to be extremely poisonous chemicals, the vapors of which are even more insidious than most poison gases because they can penetrate brick walls without giving any noticeable warning of their presence by odor. If these containers were to leak and kill persons in the buildings next door it would make very little impression on the jury to argue that the containers appeared to be in good condition when they were inspected the day before. The judge would nevertheless be likely to find that keeping of such poison gases was such an inherently dangerous action that absolute liability should be imposed without the necessity for showing that the owners failed to use reasonable care. A California case so held this finding in an instance where hydrocyanic acid gas penetrated a twelve-inch brick wall to kill a man next door.

Res Ipsa Loquitur

There are other situations where it may not be necessary to actually prove that a person was negligent in order to obtain a judgment against him for damages resulting from a fire or explosion on his premises. These are situations where the doctrine of *res ipsa loquitur* (r.i.l.) applies. The literal translation of this Latin phrase is "Let the thing speak for itself." This doctrine is applied to relieve the plaintiff of the necessity for proving negligence directly where all the factors causing the injury were under the control of the defendant and the injury would not likely have occurred if the defendant had not been negligent. To give an example, suppose the night clerk in a hotel were to discover a fire in a trash pile around midnight, and instead of immediately calling the fire department, he starts ringing all the rooms. During the fifteen minutes that elapses before the fire department is finally called a man is suffocated from heat and fire gases in his room on the third floor. Further, management had never instructed the clerk in how to sound the alarm gong nor the general procedure to follow in case of fire. From these facts alone it is possible under the r.i.l. doctrine to infer that the management was negligent and liable to the wife of the deceased guest. A court so held this finding in a Kansas case.[23]

Because everyone is presumed to use ordinary care, the burden of proof is usually on the person bringing the law suit to establish that the defendant has been negligent, i.e., has not acted as an ordinary prudent man would have done under like circumstances. This presumption that a person uses ordinary care is rebuttable, however, and when the doctrine of *res ipsa loquitur* is applied, instead of bringing in evidence to prove the defendant's negligence,

---

[23] Parker v. Kirkwood, 134 Ka. 749, 8 P.2d 340 (1932).

the plaintiff merely establishes what is called a *prima facie* case, i.e., facts surrounding the fire to show that all the factors were under the control of the defendant, and that such fires do not ordinarily happen unless someone has been negligent. If any explanation is possible at all, the defendant is the only one who has the knowledge of what happened sufficient to account for it.

A *prima facie* case is established when the plaintiff shows sufficient facts that a judgment could be rendered in his favor if no other evidence is presented by the defendant to rebut it. If there are sufficient facts from which negligence could be *inferred,* then the burden of going forward with the evidence is transferred to the defendant who must then prove that he was not negligent in order to avoid liability.

It has been held that the plaintiff made a *prima facie* case under the *res ipsa loquitur* doctrine in the following instances, to give a few examples:

1. A dehydrator tank exploded during welding operations in which there was no water in the tank that had formerly held flammable liquids. The defendant had told the welder that the tank was prepared for welding operations by washing and filling with water and by demonstrating that the tank was filled.

2. Where the plaintiff was a spectator injured in a panic at a circus having no fire extinguishers.

3. Against a warehouseman where his servant was seen using a lamp near the point of origin of the fire.

4. Against a hotel owner, where the plaintiff lost his son in the panic of a fire and threw himself through a second-story window (which would not open) falling unconscious to the ground. The hotel's halls were inadequately lighted and there was no alarm system.

The doctrine of *res ipsa loquitur* was not allowed to be applied in the following cases because of lack of exclusive control of the situation on the part of the defendant:[24]

1. A fire broke out at a gas station while the defendant was delivering gasoline, but during the time that he was doing so, several other persons also made deliveries of gasoline.

2. A gas explosion occured in a vault, but the defendant controlled only a portion of the pipes and appliances in the vault; gas could have escaped from the pipes belonging to a third party.

An unusual result was reached in a case where it was held that the doctrine of *res ipsa loquitur* was applicable notwithstanding the fact that the instrumentality was not in the defendant's possession and control. The case involved fire set to a house by a heating unit installed by defendant. The occupants of the house showed that they were in no manner negligent in its control or operaton.[25]

---

[24] *Res ipsa loquiter,* 8 A.L.R.3d 974. This annotation collects the cases in which recovery is sought for personal injury, death, or property damage allegedly caused by fire, and the owner or occupant is named defendant, and the questions of applicability of *res ipsa loquiter* is raised.

[25] Plunkett v. United Electric Service, 214 La. 145, 3 A.L.R.2d 1437, 36 So.2d 704 (1948).

Before closing the discussion of the doctrine of *res ipsa loquitur*, a doctrine which we have seen is used to transfer the burden of going forward with the evidence to the defendant, and a means whereby the plaintiff can present his case with *indirect evidence*, it would be well to look at a complication that sometimes arises, i.e., *indirect* evidence can be of two kinds: (1) inference, and (2) presumption. Both by common law and statute we have seen that it is presumed that every person uses ordinary care. A *presumption* is a deduction which the law expressly directs to be made from the particular facts. A *rebuttable presumption* may be controverted by other evidence, direct or indirect, but unless so controverted the jury is bound to find according to the presumption. The interesting question is: does the presumption of due care yield before the inference of negligence arising under the *res ipsa loquitur* doctrine? Since r.i.l. only makes out a *prima facie* case of negligence, it was held in a 1951 California case that the jury should be instructed as to both the inference and the presumption; the court said, "Nothing short of conclusive evidence which would justify a directed verdict against him can deprive such a defendant of the benefit of the presumption."

Liability for Spread of Fire

A person is liable for injuries resulting from negligence in not taking proper precautions to prevent fires or prevent their spread. This common law rule is adopted by statute in many states and provides for both criminal and civil liability. It provides that any person who, personally or through another, wilfully, negligently, or in violation of law sets fire to, allows a fire to be set to, or allows a fire kindled or attended by him to escape to the property of another (whether public or private) without exercising due diligence to control such fire is liable to the owner of such property for the damages caused by the fire. The expense of fighting such a fire is a charge against the person responsible and may be collected the same as if it were part of a written contract.

There are many cases holding that the owner of a building is liable to the owners of surrounding buildings where the fire in his building was caused by his negligence. Too often this possibility is ignored because of the common practice of owners of buildings to carry fire insurance on them. An incident occurred, known to the author, in which the tenant left the pressing iron on in her small apartment's kitchen and it burned a hole through the ironing board without setting fire to either the board or the room but smoked up the paint considerably. The tenant offered to pay the landlord the $100 estimated cost of repainting the kitchen submitted by a painting contractor, but the landlord merely collected the $500.00 offered by the insurance adjuster and refused the tenant's offer. The insurance company then collected the $500 from the protesting tenant.

Insurance codes of the various states frequently have a provision similar to that of California's which sets forth standard provisions of fire insurance policies and provides that the company shall not be liable for loss or damage occurring "while the hazard be materially increased by any means within the control of the insured." Although insurance companies do not generally avail themselves of this clause to avoid liability under the policy, yet this possibility should be considered for it is quite probable that if the insured creates a fire hazard through violation of fire regulations, his fire insurance policy would thereby be suspended and a loss proximately caused by such a violation would not be covered by the policy

In an NFPA *Quarterly* article, the assertion is made that: "It seems reasonably clear that it is incumbent upon a warehouseman and probably most commercial enterprises to install and to maintain some workable device or system for the purpose of arresting and controlling fires."[26] Cases were cited where negligence upon the part of warehouse owners and operators was predicated upon the defendant's failure to have a sprinkler system and a nightwatchman, e.g., Ricks v. Thielepape (222 S.W. 2d 399), and Tubbs v. American Transfer and Storage Company (297 S.W. 670). Other cases were cited where the failure to maintain fire doors and sprinkler systems in a working condition was the basis for imposing liability due to the spread of fire; the respective duties of the landlord and tenant are also discussed, and it is pointed out that in the absence of a contract to maintain fire protection equipment or in the absence of any actual negligence on his part, the landlord is not usually liable in such cases.

Negligence Per Se

Violation of a law or ordinance may amount to *negligence per se,* i.e., may amount to negligence even in the absence of proving facts that would show a lack of ordinary care (*per se* means "as such" or "in and of itself"). The reason for this is because the regulation itself establishes the standard of care which a person should follow. If he fails to do so it has the same effect as though the negligence were conclusively presumed. Cases involving this principle are numerous.

For example, an incinerator was located too close to the eaves of a garage and set the shingles on fire. The burning shingles set the roofs of other garages on fire all along the alley. The owner of the garage where the fire started was reimbursed by his insurance company for the damage to his garage, but the insurance companies who paid the claims of the owners of the other garages sued the owner of the incinerator and got a judgment on the grounds that the law required the stack to be at least five feet away from the wooden garage and this one was only two feet. The fact that a fire inspector had looked at the installation only a week before and had tacitly approved it by not calling the violation to the attention of the owner did not constitute a good defense. Like most policies, the insurance did not cover the damage to any of the garages but his own. The fire inspector perhaps thought he was doing the man a kindness in not requiring him to move his incinerator, and if it had been only a matter of a few inches one way or another it would probably have made no difference because its location would probably have not caused the fire. As it was, the

---

[26] Young, A. L., "Liability for the Spread of Fire, NFPA *Quarterly,* Vol. 56, No. 3, Jan. 1963, pp. 244–248.

difference in location was sufficient to create the hazard, and under the typical ordinance found in many states the owner was guilty of a misdemeanor in "permitting" the condition to exist in violation of the law. Usually, where it is a misdemeanor to create a hazard, it is also a misdemeanor to aid, suffer, abet, or permit the continuance of the hazard. Each day the hazard continues is a separate offense so as not to fall within the statute of limitations as far as prosecution of the person creating the hazard is concerned.

The fire escape cases also furnish ample examples of how liability can be imposed where a violation of an ordinance or order of the fire department results in injury to persons. In a California case it was held negligence *per se* to maintain a fire escape with a larger hatchway in it than allowed by law, and both the landlord and tenant were held liable as a result of the plaintiff's falling through the hole.[27] Another case held the owner of the property responsible since no fire escape had been provided at all as required by law. Had the tenant also been sued he undoubtedly also would have been held liable.[28] A hotel was held liable in another case for the deaths of guests who were killed in a fire where there were no fire escapes as required by law.[29]

Burning out of hours has also given rise to negligence *per se* when the fire spread to adjoining property. When the persons attending a rubbish fire in the City of Vernon, California, went off and left some glowing embers at 9:00 AM (the time when legal burning of trash was supposed to end), the court said: "Under the ordinance it was clearly the duty of the defendant to regulate the magnitude of any fire it contemplated starting so as to control and extinguish it within the time limit prescribed by law. Having failed to do so this was negligence, and this negligence not only contributed to the burning of the plaintiff's property, but the failure to comply with the mandates of the law is the primary, if not the sole reason, for the destruction thereof."[30]

California state law makes it unlawful to burn trash in an incinerator having openings in the spark arrestor over one-fourth inch square, or which is located within ten feet of a building. A decision held that it was negligence *per se* on the part of a tenant to do so. The landlord who rented the premises with the unlawful incinerator was guilty of contributory negligence that barred his insurance company from recovering from the tenant the amount of the loss claim paid the landlord.[31]

In a case, growing out of a restaurant fire in Seattle, Wash., an inspector had apparently overlooked an illegal grease duct connection between the hood over the range and the general ventilation system for the building. The system did not have the insulation or clearances from combustible framework as required for regular grease ducts. Notwithstanding the fact that the duct had been installed many years before the defendant took over the premises, or the fact that the fire marshal had been in and out of the place many times without ever noticing the partially obscured illegal connection, the federal

---

[27] Mars v. Whistler, 49 Cal. App. 364, 193 P. 600 (1920).
[28] Roxas v. Cogna, 41 Cal. App.2d 234, 77 A.L.R. (1940).
[29] Hoopes v. Creighton, 100 Neb. 510, 160 N.W. 742 (1916).
[30] Alechoff v. Los Angeles Gas and Electric Corp., 84 Cal. App. 33, 257 P. 569, (1927).
[31] Travelers Indemnity v. Titus, 265 Cal. App.2d 515, *Cal. Rptr.* 490 (1968).

court sustained the jury's decision in the trial court against the restaurant company for damages to the property of others through which the fire in the grease duct had spread. Hence, this case illustrates how even an unintentional violation of the law may result in liability, and how the fire inspector can be of real assistance to the property owner in helping him detect and correct any conditions which may give rise to a fire and subsequent lawsuits.

Another case held that the tenant was liable for damages to his landlord's warehouse because he had piled his bales of rags in such a manner as to so completely block the aisles and doors as to effectively obstruct the hose streams of the fire department and access by fire fighters.

Personal liability, then, can be an effective talking point for any fire inspector who wants to supplement police power with the power of persuasion. The person who may not be concerned with the possibility of "notices" being served upon him, or even a small fine for disobedience of an official order, may nevertheless be financially interested in the knowledge that a failure to maintain his safety appliances in proper condition can render him personally liable to anyone injured as a result thereof.

To the man who feels a smug disinterest toward any suggestion of increased firesafety due to the fact that he has fire insurance coverage, the questions should be asked: "Does your fire insurance cover losses sustained by your neighbors when, through your failure to heed these recommendations, a fire breaks out and spreads to their property, and you are subsequently held liable for your negligence? Does your insurance cover the liability to persons who may be injured in an effort to leave your building through doors that swing the wrong way or by fire escapes that fail to function, when, in the absence of fire, they are fleeing from the building because of earthquake or panic? Does it cover the liability to persons who are injured while fighting the fire in defense of their property after it has spread from your premises?"

An individual's possible criminal liability also should not go unstressed, for, as is illustrated by the infamous Coconut Grove night club fire that occurred on November 28, 1942, in Boston where the owner and manager (among others) were indicted on manslaughter charges (following the death of almost 500 patrons), it becomes obvious that the disregard of fire laws may result in more than damage suits or the paying of fines.

Ordinances

Another method which a fire prevention bureau may employ to promote firesafety, and perhaps the most common, is the enforcement of fire ordinances, with the filing of complaints and imposition of fines or other penalties as an incident thereto.

Notwithstanding the fact that much can be accomplished through sales technique applied to fire prevention, even in the absence of a finely drawn fire prevention code, it is an invaluable asset to the fire prevention engineer to have a good set of ordinances with which he can back up his safety arguments.

One of the most difficult duties imposed upon a fire prevention bureau is the drafting of fire ordinances for submission to the local legislative body for enactment. It is not enough to merely draw up a "catch-all" ordinance, which in effect forbids the creation or maintenance of a fire hazard, for immediately the question is asked: "What is a 'fire hazard?'" In a Montana case where a statute authorized the fire marshal to condemn buildings constituting a fire hazard, the court said that it was not necessary that the danger or risk be immediate; the word "hazard" in this connection meaning risk, danger or peril. It was contended that since forty feet intervened between the dilapidated building which had been ordered removed or repaired, that no "fire hazard" existed; the court said, "We are given no measuring stick for the detemination of what is a fire hazard other than the definition of a 'public nuisance' found in the statute itself."[32]

Careful consideration should be given to content of proposed ordinances. The draftsman should have sufficient farsightedness to consider the entire field instead of the particular problem facing him at the moment. To do as many are inclined to do, i.e., write a new ordinance any time the ones at hand do not adequately handle the present situation, will result in a disjointed compilation of "spur-of-the-moment" ordinances. Those relative to fire prevention should be revised as a body, rather than as individual units, if proper relationship between the various subdivisions is to be maintained.

Reasonable underlying principles of firesafety should be the basis of every law enacted, so that in applying these ordinances, it becomes unnecessary to quote the law as authority for making the particular recommendation.

A happy medium should be reached between legislating in too much detail and too generally, and the modern tendency is toward less detailed regulation. General regulations, patterned after those recommended by the National Fire Protection Association, make for greater uniformity between neighboring cities, and call for less frequent revision.

A satisfactory way of handling details, in states where local governments are permitted to do so (by enabling acts of the state legislature), is to incorporate by reference the latest national standards of various organizations which draw up technical requirements for the different phases of fire safety, e.g., the American Petroleum Institute, the American Society for Testing and Materials (ASTM), the National Fire Protection Association, etc. A number of states have passed such acts. In California the court declared that if standards are adopted by reference, e.g., the National Electrical Code, a specific edition should be cited.[33]

The common practice of states and cities in drafting fire prevention legislation has been to include all detailed requirements. This practice is not calculated to keep such legislation in step with new inventions and the development of industry, as the machinery for extensive revision is too difficult to put into motion with the necessary frequency.

---

[32] Brooks v. Cook, 84 Mont. 478, 276 P. 958 (1929).
[33] Agnew v. City of Culver City, 147 Cal. App.2d 144, 304 P.2d 788 (1956). Affirmed, IDEM, 51 Cal.2d 474, 334 P.2d 571 (1959).

Another thing which must be considered in drafting municipal ordinances is the problem of how to adopt local regulations which will not duplicate nor conflict with state laws on the same subject.

A more perplexing question is whether a municipality can impose penalties for acts which are likewise declared to be a crime under the laws of the state, and if so, whether the prosecution under the ordinance may be pleaded in bar to a prosecution under the statute. For example, can a person who has been arrested for selling firecrackers in violation of a city ordinance be imprisoned for violating a state law to the same effect?

Before 1850 it was generally assumed that in absence of express authorization, an ordinance attempting to punish acts already criminal under the laws of the state was void. "Although New England still clings to this doctrine, in most of the other states opposition to coordinate legislation has been broken down completely."[34]

In the period arond 1868, the majority view was that when both an ordinance and a state law prescribe a penalty for a given act a conviction or acquittal under one was a bar to a prosecution under the other.

Now, however, it is well settled that the same act may constitute an offense against both the state and the municipality and may be punished by either or both. There are various reasons advanced for this rule, but an increasingly popular one is based on the analogy to the similar rule sustaining successive state and federal prosecution. "We have erected our doctrines of civil crimes, coordinate laws and successive prosecutions. But the California lawyer who attempts to act upon the basis of either of these doctrines themselves or of the supposed principles upon which they rest goes far astray. Not one of the rules which are commonly said to be 'definitely established' is applicable here. Nor is there another state of the Union in which the same rules are in force."[35] The author of the cited reference points out that whereas in New England this rule would seem to rest upon the theory that only an express statutory authorization could validate such an ordinance, in California it rests upon the broader ground that the constitution itself forbids coordinate laws.

In California the court said on the question of the validity of a municipal ordinance punishing acts already criminal under the laws of the state that "... it would seem that an ordinance must be conflicting with the general law which may operate to prevent a prosecution of the offense under the general law. The Constitution provides that no one shall be twice put in jeopardy for the same offense. If tried and convicted or acquitted under the ordinance, he could not be again tried for the same offense under the general law."[36]

Thus, unlike most states, which sustain both coordinate legislation and successive prosecutions, California has almost consistently followed the above opinion in holding such ordinances invalid. However, the court has been careful to point out "that we only hold that there is a conflict where the ordinance and general laws punish precisely the same acts. We do not wish to be under-

---

[34] "Penal Ordinances in California," 24 Cal. L. Rev. 123, 128 (1936).
[35] *Ibid.,* p. 128.
[36] *In re Sic,* 73 Cal. 142, 148, 14 P. 405, 408 (1887).

stood as holding that the sections of the ordinance which make criminal other acts not punishable under the general law are void because the legislature has seen fit to legislate upon the same subject."[37]

The court has been very liberal in interpreting city ordinances as not conflicting with state regulations—e.g., in *Ex parte* Boswell,[38] the court held that a statute prohibiting gambling does not prevent a municipal corporation from making it a misdemeanor to visit a place for the purpose of gambling.

Later a further rule was propounded in California that "the only way the legislature can inhibit local legislative bodies from enacting rules and police regulations is by the state occupying the same legislative field so completely that legislation on the subject by local legislative bodies will necessarily be inconsistent with the state act."[39]

Hence, in California, one is left with the perplexing problem of "where has the state law completely occupied the field?" J.A.C. Grant, in his exhaustive research of California precedents in this field, says by way of summary:

"Although random precedents may be found for contrary rules, the prevailing doctrine holds that a penal ordinance is valid unless it punishes precisely the same acts which are criminal under the laws of the state or conflicts with the carrying out of the policy of the state law. When acting under the powers derived from the court, even an express statutory provision cannot narrow the scope of local legislation beyond this point. Even an ordinance so phrased as to include all or part of the analogous offenses embraced in the state law is valid as to the excess and a complaint under it need not state facts sufficient to negate the commission of a state crime—this being a matter of defense. No matter how similar the acts dealt with in the statute and ordinance may be, the local government has complete legislative discretion, subject only to the outer limits set by its charter or by the constitution, to provide whatever penalties it sees fit for those violating the ordinances. These penalties may be less than, equal to, or greater than those provided by statute for similar or virtually identical acts."[40]

Safety Standards

It is not necessary that there be a violation of an ordinance or statute to establish liability for the spread of fire. National standards, such as those promulgated by the NFPA, can have an important bearing on the question of negligence in the provision of adequate fire protection, as is well illustrated in the case of an action brought under maritime and admiralty jurisdiction against a marina to recover damages by fire to a yacht in winter storage. The court held that the marina, "as bailee of the yacht for winter storage, took none of the precautions required by Fire Protection Standard for Marinas and Boatyards (NFPA No. 303) published by the National Fire Protection Association, employed no watchman, and provided no water source . . . the marina

---

[37] *Ibid., In re* Mingo, 190 Cal. 769, 214 P. 850 (1923).
[38] 86 Cal. 232, 24 P. 1060 (1890).
[39] *In re* Iverson, 199 Cal. 582, 250 P. 681 (1926).
[40] "Municipal Ordinances Supplementing Criminal Laws," 9 S. Cal. L. Rev. 95.

was negligent and its negligence was the proximate cause of the spread of fire (from point of origin) to the nearby yacht, notwithstanding that neither party had information revealing origin of the fire." Judgment was rendered in this case for the plaintiff.[41]

Other Methods of Promoting Fire Safety

In addition to filing complaints for violations of ordinances, abating nuisances, and stressing personal liability, there are other available legal procedures often employed to promote firesafety.

Some departments require that bonds be posted, which may be forfeited upon the breach of lawful conduct in the operation of a business granted a permit. The proceeds upon such a breach can be used to defray the expense of putting the property in a lawful condition. Such is the practice in Los Angeles with respect to oil wells, refineries, and natural gas plants, wherein a bond is required with the application for a permit for their maintenance or operation.

Penalty actions, similar to the old common law *Qui Tam,* where an action is brought under a law imposing a penalty for doing or not doing an act, and allowing part of the penalty to go to the plaintiff suing for the same and part to the state, may also be employed by the fire prevention bureau.[42]

Licensing, with threat of revocation of the license, is another effective control in the field of fire prevention. The exacting of a license which interferes with interstate commerce, while invalid, does not render the city liable. If the occupation is harmful it is said to be a license fee that is imposed; but if it is innocent, then it is a license tax. The fire department, however, need only employ the license as a device to regulate those businesses which affect the public safety, and hence the fee imposed should be no more than that which is necessary to defray the expenses of inspection and clerical work. The courts, however, are not too exact in determining the cost of regulation and supervision, yet any amount disproportionate to the expense of issuing the license and regulating the business would be unreasonable and invalid. A failure to obtain a license from the city may violate a state law.

Conflict of Jurisdiction

Before leaving the subject of fire prevention laws and their enforcement, it might be well to consider the problems that not infrequently arise with respect to multiple and conflicting requirements of various political agencies on a given subject.

It has already been noted that enforcement of municipal ordinances which do not conflict with, nor regulate in exactly the same manner as, state laws is proper, and, of course, the state laws themselves can be enforced by any

---

[41] Fireman's Fund American Insurance Company v. Captain Fowler's Marina, Inc. 343 Fed. Supp. 347 (1972).
[42] Boston and Maine R.R. v. Armburg, 285 U.S. 234, 52 S. Ct. 336 (1932).

peace officer. Even a private citizen may arrest another for a public offense committed in his presence when authorized by statute or for a breach of the peace at common law, and such public offense, breach of the peace, or other criminal act is not limited to mere violations of local ordinances. And it should be remembered that any citizen can always file a complaint against the violator of any law, whether it be local, state or federal.

Penal laws are generally said to have a territorial limitation, so as to preclude, for example, a Los Angeles fire inspector from arresting a Pasadena resident for a violation of a Los Angeles fire ordinance (since his powers extend only to the area in which he is an officer). Yet it cannot be said that an officer is powerless to enforce all the relevant laws that are applicable to the citizens within the political subdivision of which he is an officer.

Where there is a conflict between a municipal ordinance and a county ordinance, the city ordinance has superior force within the municipal limits. But where a defendant was doing business authorized and in conformance to a municipal ordinance, but without a license from the county which was imposed for revenue purposes, he was convicted for violation of the California Penal Code, which makes it a misdemeanor to carry on any business without a license required by law.

As an example of where possible conflicts can arise, let us consider the problems which must be confronted in regulating dangerous chemicals.

Federal Regulation

The federal power to regulate explosives and dangerous substances is derived from the Commerce Clause of the Constitution. On purely local matters, not involving or affecting interstate or foreign commerce, the federal government has no regulatory power, no matter how dangerous the article may be; in such instances, the police power is vested solely in the states and local governmental bodies.[42]

However, most of the serious problems affecting the handling of hazardous materials will arise in the field of interstate and foreign commerce. At the present time, the chief dangers arise from its transportation by air, ship, or rail, and this is clearly within the realm of federal regulation.

The question then arises as to how far the federal government can go in regulating the shipment and transportation of dangerous substances, such as ammonium nitrate, hydrocyanic acid, liquid oxygen, fluorine, etc. Under the Commerce Clause of the Constitution, the power of Congress over interstate and foreign commerce is practically absolute, and may extend even so far as to exclude an article from commerce entirely.[43]

This very extensive power over interstate and foreign commerce, however, is not completely unrestricted, and the regulatory power with respect to foreign commerce may be broader than that pertaining to interstate commerce. Moreover, the power to regulate commerce does not carry with it the right to destroy those limitations and guarantees which are contained in other provisions of

---

[43] Brolan v. United States, 236 U.S. 216, 35 S. Ct. 285 (1915).

the Constitution. The test as to the limits of Congressional power has been well established since the case of McCulloch v. Maryland,[44] in which, as emphasized in Interstate Commerce Commission v. Brimson,[45] the Supreme Court said:

"The sound construction of the Constitution must allow to the national Legislature that discretion, with respect to the means by which the power it confers are to be carried into execution, which will enable that body to perform the high duties assigned to it, in the manner most beneficial to the people. Let the end be legitimate, let it be within the scope of the Constitution, and all means which are appropriate, which are plainly adapted to that end, which are not prohibited, but consistent with the letter and spirit of the Constitution, are constitutional."

Thus, a safety regulation promulgated by a federal agency, under an Act of Congress, must not be arbitrary and capricious. It must be consistent with the hazards involved, and if experience has shown that the transportation involves serious dangers, the regulation may be quite comprehensive; it is evident that dangerous chemicals must be considered as falling within that class of hazards.

State and Local Regulations

Local governmental authorities can be expected to regulate the dangers created in their jurisdictions by the presence of dangerous chemicals. Since the safety problem usually arises in congested and built-up areas, the regulatory measures will usually be promulgated by such public entities as municipalities, port authorities, etc., acting pursuant to a lawful delegation of power from the state. Where an unincorporated area is involved, the state itself may enact such regulatory measures as it considers necessary.

Aside from the possible conflict with the interstate commerce power of the federal government, the power of local governments to protect the safety of their citizens by regulatory measures is far-reaching. The exercise of police power for the safety, health, and general welfare of its citizens is inherent in every state government, and, through it, in every municipality and organized local government.

With reference to the exercise of police power by a state, the standard test of reasonableness is consistently applied to determine its validity. This same test would be applied in the case of regulatory measures controlling dangerous substances, such as chlorine, blasting agents, and ammonium nitrate. In a study prepared by Attorneys Steptoe and Johnson, of Charleston, W. V., shortly after the Texas City explosion in 1948, the following comments were made:

"To what extreme limits local regulation may go, therefore, depends upon the particular circumstances involved. Safety measures adopted by local governments will not be overthrown by the courts, unless they clearly go so far as to be classified as arbitrary or unreasonable.

---

[44] 17 U.S. 579, 4 Wheat. 314 (1819).
[45] 154 U.S. 447, 14 S. Ct. 1125 (1894).

"For example, if the municipality of Texas City should now adopt an ordinance requiring ammonium nitrate to be loaded and unloaded in isolated parts of the city, or even beyond the city limits, it seems unlikely that a court would say that such safeguards were unnecessary and unreasonable, in the light of the holocaust which Texas City so recently experienced.

"It is to be noted, too, that states and local governments have the power to adopt and enforce safety regulations applicable to ammonium nitrate and other dangerous substances in a field where the Federal Government cannot enter at all—i.e., with regard to the manufacture, storage, use, and handling of the dangerous article, where interstate commerce is in no way involved. In that field of local police regulations, the power of the states and local governments is supreme and exclusive. For example, a state could enact a statute, or a municipality an ordinance, prescribing reasonable safeguards for storage and use of ammonium nitrate within their boundaries, *before* any shipment of the article in interstate commerce was begun, or *after* its travels in such commerce came to an end."[46]

Another example cited in the above study of what local governments may do in the interest of safety is the extent to which they have, over the years, succeeded in practically banning fireworks from our national life. "Everyone can remember when the display of fireworks was regarded as almost a necessity for the proper observance of the Fourth of July. Yet, through the years, the tragedies experienced in the handling of fireworks, especially among children, actuated the local governments to forbid the sale or display of fireworks within their boundaries. Thus, the widespread use of a hazardous article of commerce has had to give way to safety measures imposed by the local governments."[47]

At the present time Congress is drafting legislation that will provide federal regulation of fireworks beyond that currently provided by the Department of Transportation for their shipment in interstate commerce.

Conflict Between Federal and Local Government Regulations

In seeking to adopt safeguards for the handling, shipment, and transportation of dangerous chemicals, both the federal government and local agencies can adopt their own measures, and these may sometimes have conflicting requirements. The regulation adopted by the local government may be more restrictive than the federal measure, or vice versa, and where this occurs, the shipper must comply with both insofar as possible.

The federal government has the primary power of regulation of foreign commerce and interstate commerce under the Federal Constitution. But it is not correct to say that it may entirely preempt the right of local agencies to impose any regulations for the protection of their local citizens. Where a shipload of dynamite had exploded in Seattle, Wash., harbor, the question involved was whether the City of Seattle had the right to specify a particular place in the

---
[46] Steptoe and Johnson.
[47] *Ibid.*

harbor where ships carrying explosives in interstate commerce might be anchored. The court said, in answer to the argument advanced that only the federal government could impose such a regulation: "It is beyond question that, where Congress has legislated in respect to either foreign or interstate commerce, no state or other subordinate legislation upon the same subject is of any validity. But we find no legislation of Congress with respect to the place or places within any harbor of the United States where any kind of explosives shall be handled, kept, or stored. . . . It is beyond question that where Congress has legislated in respect to either foreign or interstate commerce no state or other subordinate legislation upon the same subject is of any validity, yet we understand the law to be that, where Congress is silent, the state may legislate in aid of, but without burdening, both foreign and interstate commerce."[48]

Thus it would appear that local regulations, except in a field of national concern requiring uniformity, are valid even though they may indirectly affect interstate or foreign commerce, so long as they do not impede the free flow of commerce and have not been made the subject of regulation by Congress. In determining whether local regulations have been superseded by federal action, it is held that they are so superseded only when the repugnance or conflict is so direct and positive that the two acts cannot be reconciled or consistently stand together.[49]

The states do not have the power, however, in the exercise of their authority to regulate matters of local concern to exclude, directly or indirectly, the subjects of interstate commerce.[50]

Regulation of Blasting Agents, as an Example of Conflict

A good example of such a conflict arose in the case of E. I. duPont de Nemours & Co. v. the Board of Standards and Appeals of New York. The Board of Standards had affirmed a determination of the fire commissioner that "Nitramon," a blasting agent manufactured and shipped by du Pont, was an explosive within the purview of the applicable provisions of the Administrative Code of New York and therefore could not be handled across any pier in the city in excess of 2,500 pounds. The code required that all such shipments in excess of 2,500 pounds and not exceeding 5,000 pounds, must be transferred from vessel to vessel at a distance of not less than 1,000 feet from any pier line. Thus, du Pont under this ruling was entirely prohibited from loading Nitramon from city piers onto vessels though intended for foreign shipment.

In holding that this restriction was invalid as to interstate and foreign commerce, the court said:

"The question here presented involves a consideration of the statutes and regulations of the federal government governing such shipment in interstate and foreign commerce as well as the provisions of the local law, keeping in mind the basic principle that in this field, if federal action has been taken, local law must be deemed inoperative if inconsistent or conflicting in any respect

---

[48] City of Seattle v. Lloyd's Plate Glass Insurance Company, 253 F. 321 (9th Cir. 1918).
[49] Kelly v. State of Washington, 302 U.S. 1, 58 S. Ct. 87 (1937).
[50] Baldwin v. Seelig, 295 U.S. 511, 55 S. Ct. 497 (1935).

therewith. Federal regulation supersedes a local law on the same subject, even though it involves a matter of the police power of government, if the field of operation is under our constitution in the federal domain (Quaker Oats Co. v. City of N.Y., 295 N.Y., 527). The Federal Government, it should be observed, is no less zealous of the public interest than the local authorities."

The City of New York argued that its local law occupied a field not covered or preempted by federal law, was not inconsistent with it, and did not violate the Commerce Clause of the Constitution, although it may have incidentally affected a matter of interstate and foreign trade. They contended that the rule to be applied is that when Congress occupies only a limited field, local law in the proper exercise of the police power outside that limited field is not forbidden. In this case the federal law consisted of a limited number of regulations regarding the packaging of nitrocarbonitrate (the product which du Pont tradenamed "Nitramon") for transportation in interstate and foreign trade, while the local law merely incidentally regulated its handling as an explosive in the territorial limits of the City of New York.

The court acknowledged that the local law expressly stated that nothing contained in the law should be construed as applying to the transportation of any article or thing shipped in conformity with the regulations prescribed by the Interstate Commerce Commission, nor be construed as preventing the enforcement of reasonable local regulations not inconsistent with or in conflict with the regulations of the Commandant of the Coast Guard.

Title 46, United States Code (sec. 170, subdiv. 7 [a]) provides that "in order to secure effective provisions against the hazards of health, life, limb or property created by explosives or other dangerous articles or substances," the Commandant of the Coast Guard shall by regulations define, describe, name and classify all explosives and other dangerous articles.

The court also recognized that "Nitramon" is a blasting agent developed by du Pont to replace commercial dynamites, and when caused to explode will detonate to produce an explosion equivalent to that produced by large diameter dynamite. However, the court thought that because this blasting agent required a primer to detonate it, that therefore it was different from an explosive. It said: "In other words, 'Nitramon' by itself is not truly an explosive, but may be detonated by an explosion in its immediate area." The city, however, pointed to the fact that Nitramon is composed principally of ammonium nitrate (92.42 percent) and thus came clearly within the definition of an explosive in the code and that a serious shock or intense fire could cause it to explode and, when involved in a fire, it may detonate. Hence, the city argued that it should be subjected to the restrictions applicable to all explosives set forth in the code in order to avoid possibility that it might be detonated and cause serious injury to the people and property of this congested area.

To this the Court replied:

"Although, under the one broad definition in the Code, Nitramon was determined by the Fire Commissioner to be an explosive, thus requiring enforcement at city piers of the Code restrictions with respect to its handling and loading to the same degree as any other explosive, the Interstate Commerce

Commission as well as the Commandant of the Coast Guard have classified it as an oxidizing material rather than a direct explosive. The federal classification divides explosives into three groups, A, B and C, and other dangerous articles into separate categories, one of which is oxidizing materials. In 1954 the Coast Guard regulations made a further refinement in classification of ammonium nitrate, the equivalent of Nitramon, depending upon the type of coating and packaging, and established specific rules for loading and discharging of each group.

"The Coast Guard regulations reveal a comprehensive, detailed and complete system of regulation and supervision. The captain of the port may supervise and control the transportation, handling, loading, discharging, stowage or storage of explosives and other dangerous articles or cargo (33 Code of Federal Regulations, sec. 6.12-1). The Commandant may designate waterfront facilities for the handling and storage of all such articles and for the loading and discharge thereof from vessels, and may require owners of vessels to obtain permits for loading and unloading from the captain of the port (33 Code of Federal Regulations, sec. 6.12-3). A waterfront facility is defined as a pier, wharf, dock or similar structure (sec. 6.04-4). Only such facilities as fulfill the numerous and exacting safety requirements of section 126.15 may be so designated. There are stated maximums for each class of explosive and for each category of other dangerous articles beyond which they may not be handled, etc., at any one time without notification to the captain of the port: the permissible maximum for oxidizing materials is 100 tons (sec. 126.27 [b]). The captain of the port is given authority to require any transaction involving any class or category in any amount to be conducted under his supervision (sec. 126.29) and to terminate or suspend the permission given by the regulations whenever he deems such action necessary for the security or safety of the port, vessels or waterfront facility (sec. 126.31).

"With regard to ammonium nitrate, there has been a change in the regulations. Prior to the 1954 reclassification it was provided that shipments of this product in amounts exceeding 500 pounds could not be loaded on or discharged from any vessel until authorized by the district commander, who was to withhold permission if this was to be done in a congested area of the port until a more remote area was available (46 Code of Federal Regulations, sec. 146.22-30). The 1954 amendment revised this regulation by providing different requirements for each of the types of ammonium nitrate now established; for example, when organic coated and packaged in paper bags, it could not be loaded or discharged except at facilities so remotely situated from populous area and/or high value or high hazard industrial facilities that loss of lives or property would be minimized in case of explosion or fire. Nitramon comes within the far safer classification of (c) (2), i.e., of dynamite grade, without organic coating, and packaged in sealed metal containers. It is there provided that it may be loaded or discharged at any waterfront facility which conforms to port security and local regulations and that no permit is required for this transaction. (It will be observed that it is only the pier where the loading is to take place and not the loading itself which shall conform to Coast Guard as well as local regulations. But this does point the way to a practical solution of the problem. Arrangements may be made by the city authorities with the Coast Guard to have this product loaded at selected piers.)

"It is abundantly clear that the federal government has acted and that federal regulations do cover the specific situation here involved—the loading and unloading of Nitramon on the city's piers and docks for shipment in interstate or foreign commerce. The loading and unloading of cargo is one of the acts in the shipping process and an essential part of the transaction in trade or commerce.

"The determination that Nitramon as an explosive under the (City) Code definition is subject to the provisions of section C 19-32.0 (d) of the Code prohibiting entirely its being landed or placed upon any city pier or dock and restricting the quantity thereof allowed to be transferred from one vessel to another in the nearby area is in obvious conflict with the federal regulations directly authorizing loading and unloading of that product up to a stated maximum amount at piers and docks which conform to port security and local regulations. To the extent, then, that the shipment is intended for interstate or foreign commerce, this determination must be held to be inoperative, since the transaction is governed by the applicable federal regulations. To the extent however, that interstate or foreign commerce is not affected, it is doubtful that the city authorities may treat Nitramon as an explosive even under the Code definition.

"The decision of the board is accordingly reversed and the determination that Nitramon, when shipped in interstate or foreign commerce, may not be loaded or unloaded on docks or piers in the City of New York or transferred from vessel to vessel except in the limited quantities set forth in the Code, being inconsistent with federal regulations covering that subject, is annulled."[51]

Notwithstanding the preceding decision, with respect to explosives and other dangerous substances, such as ammonium nitrate, it would seem that Congress has expressly recognized the right of local governments to adopt reasonable safeguards, for the Act of October 9, 1940, relating to shipping, provides: "Nothing contained in this section shall be construed as preventing the enforcement of reasonable local regulations now in effect or hereafter adopted, which are not inconsistent or in conflict with this section or the regulations of the Secretary of Commerce established hereunder."[52]

Therefore, as the conflict of jurisdictions is concerned, it is believed that local governments will be permitted considerable discretion in adopting reasonable and necessary safeguards to protect their citizens from the serious explosions. Only when they go to extremes, so that the regulation in question is more prohibitive than what the courts feel are the safety factors actually required, will the measures adopted by the local governments be overthrown.

Local Codes

The common practices of governing certain hazardous occupancies by the

---

[51] El DuPont de Nemours Co. v. Board of Standards and Appeals of the City of New York, 158 N.Y.S.2d 456, 5 Misc.2d 100 (1956).
[52] 46 U.S.C. 17, Section (7) (d).

fire department issuing permits gives rise to some interesting questions. Should a fire inspector who discovers violations of building ordinances, while inspecting for violations of fire ordinances, refuse to grant approval for the issuance of a permit by the fire department, where he finds that such violations render the building unsafe for the use applied for? Notwithstanding the fact that the enforcement of building ordinances may have been given over to the building department exclusively, the fire department should withhold approval of the issuance of a permit until the matter has been referred to the proper department for investigation. The department cannot approve an unlawful condition, even though the condition is not a matter affecting firesafety, and if it does do so, the approval is a nullity.[53]

The better practice for the fire inspector, in surveying the premises for which a permit is sought, is to render a complete written report of all the conditions which he thinks should be brought to the attention of his superiors; hence there is no necessity for the individual inspector assuming the responsibility of deciding whether the presence of certain conditions should preclude the applicant from obtaining a permit. It should not be left to the inspector to approve or disapprove, but merely to investigate and report, and in such report he should include a statement of all dangerous or defective conditions that come to his attention so that his superiors may determine the advisability of granting such a permit.

There should be close cooperation between various inspection agencies, not only within the city government itself, but also between the city and county fire prevention bureaus, and the state fire marshal's office.

It is a pleasant commentary on modern fire prevention bureaus to note the increasing tendency toward joint inspection tours on the part of city and state fire inspectors, as well as city and county fire inspectors. By each refusing to issue permits until the recommendations of the others have been complied with, the applicant is impelled to hasten his conformance. Needless to say, the effect on the applicant is salutary, for he is able to determine at one time all the acts that must be performed as a condition to receiving his permit.

Fire Inspector's Liability

What conduct will render the fire inspector liable? It has already been seen that many jurisdictions hold a fire fighter to be a public officer. Where fire inspectors are considered public officers, their tort liability is governed by the principles applicable to the latter.

A public officer may be liable the same as any other person for his torts, but while in the performance of his public duty, this liability may be limited because of privilege. Where a duty or act is discretionary, the officer is ordinarily not liable for a mistake in judgment, when acting within the scope of his employment. But where he is assigned a task which involves no act of discretion,

---

[53] Maguire v. Reardon, 41 Cal. App. 596, 183 P. 303, 255 U.S. 271 (1919).

the officer is liable for failure to perform his duty, or for acts in excess of his authority.

With respect to undue severity, discrimination, and malice, Willoughby in his book on Constitutional Law of the U.S., observed: "In general no officer is held responsible in damages to an individual for nonperformance or negligent performance of duties of a purely public or political character."[54] Certainly the fire inspector's duties are of a public character. Insofar as he is required to determine whether certain conditions meet the requirements of the law for the issuance of a permit, his duty is discretionary, and in the absence of a gross abuse of discretion, his decision, or failure to give one, would not give rise to a cause of action.

Criminal responsibility may be imposed, however, on fire officials whose breach of duty contributes to disaster. Following the Cocoanut Grove night club fire in 1942, a lieutenant of the Boston Fire Department was indicted for neglect of duty, while an inspector was indicted for failure to report violations of building laws and insufficient exits; both were acquitted. In 1963, the Indiana State Fire Marshal and the Indianapolis Fire Chief were indicted for failure to carry out their inspection duties, following the disastrous explosion in the Indianapolis Coliseum. Charges were later dropped against the chief, and the state fire marshal was prosecuted, though not punished.

An editorial in *The Trumpet* of the Uniformed Fire Officers Association in New York commented on the Indianapolis tragedy as follows: "What has this got to do with Fire Officers in New York City? Nothing, except to remind you of something you may have recently overlooked. Our obligation to inspect buildings is not satisfied by a superficial once over. This might satisfy those in Headquarters who do not sign the inspection reports. Those with superior skills rise to higher ranks. We don't know why a Fireman's inspectional ability should be equated with a Captain's and we don't know why a Captain's should be equated with that of a Deputy Chief, nor a Deputy Chief's with an Assistant Chief of Department. So many factors enter into our Fire Prevention inspections that what is true in one building may not be true in an identical physical structure next door. How much law are we required to know? The lawyers have law libraries to rely on. We don't.

"The explosion in Indiana should remind you that you, and ONLY YOU! are responsible for the inspection which you said you made. When you sign the building record card, or make out the report form, you are attesting to the fact that you made an inspection and that you made it properly. A Fire Officer is not required to be a walking law library but he does know what he is required to know to perform his functions properly. No one can excuse you from failure to do your inspectional work according to the law. Neither Officer can excuse you from doing your inspectional work according to law. If an accident takes place the first action taken will be to see who inspected the premises last. No one has any weight with a Grand Jury."

---

[54] Willoughby, *Constitutional Law of the United States,* 2nd Student Ed., Voorhis, Baker & Co., 1933, p. 605.

What is the liability of an officer who acts under an unconstitutional law? Although there are decisions to the contrary, it has been held that an unconstitutional law must be considered to have the force of law until declared void, so far as to protect an officer acting under it.[55] In some states, as in California, a statute exempts public officers and employees acting in good faith under a statute from liability in the event that the law is subsequently declared unconstitutional. (See Appendix C.)

Unless a law is clearly arbitrary, unreasonable or discriminatory on its face, the fire inspector who enforces it in good faith may feel reasonably immune from liability.

With respect to the liability of officers abating nuisances, Willoughby says: "It is established that the legislature may delegate to the municipal or other administrative authorities the power to declare that certain things are nuisances and summarily abate them as such, with or without notice and hearing. However, in those cases in which abatement is had without a previous determination by some due process of law that a nuisance in fact exists the officer acts at his peril."[56]

Though it is said to be well settled that an officer acting in good faith and with reasonable grounds for believing, though mistakenly, that a nuisance in fact exists, has a good defense, yet he must be careful lest a jury find that his action was not justified. In a Wisconsin case the jury held the officer liable, though the average man would believe a nuisance did exist.[57] In a Pennsylvania case, a grand jury found that a shanty-town was a nuisance, so the judge ordered the mayor to tear it down, and he did. The Pennsylvania Supreme Court said the mayor would not be liable only if the jury found that the buildings were in fact a nuisance.[58]

"Reasonableness" should be the test of the inspector's conduct. Although an ordinance may declare it unlawful to permit a fuse in excess of fifteen amperes in a lighting circuit, and another ordinance declares that the violation of any ordinance shall be deemed to create a nuisance which may be abated summarily, no one would contend that a fire inspector is justified in removing a twenty ampere fuse from a lighting circuit in which an electric refrigerator is connected, with resultant damage to the food therein, where he can merely notify the occupant to purchase the proper size fuse and replace the unlawful one.

On the other hand, would it not be a reasonable fire protection measure to remove a glowing-hot bridged fuse in order to prevent an immediate threat of fire to the premises?

As in the case of a fire chief who orders a building blown up to prevent the spread of fire, an emergency may give rise to the exercise of power to destroy or confiscate property, for which a mistake of judgment will not render the officer liable.[59]

---

[55] Nagel v. Bosworth, 148 Ky. 807, 147 S.W. 940 (1912).
[56] See note 54.
[57] Lowe v. Conroy, 120 Wis. 151, 97 N.W. 942 (1904).
[58] Fields v. Stokely, 99 Pa. 306, 44 Am. Rep. 109 (1882).
[59] Surroco v. Geary, 3 Cal. 69 (1853).

"Reasonableness" is also a limitation on his power of inspection. Could it be argued that a fire inspector has the right to enter a person's home unannounced, and proceed to look through all the drawers and cupboards for fire hazards? No, even where an ordinance authorizes him to inspect any place at any time.

Inspection of Private Dwellings

There is no doubt that a right to inspect a private dwelling would be considered a reasonable exercise of the police power where there is reason to believe that conditions exist which render the property a fire menace to the community; however, a reasonable notice should be given that an inspection is about to be made (unless an emergency justifies immediate entry, in which case no advance notice or search warrant is necessary). But where a residence is so isolated from other occupancies or properties so that it would render no hazard, even if on fire, the right of privacy for the private dwelling's owners might be considered paramount to the police power.

In Mapp v. Ohio,[60] the Supreme Court clearly imposed a strict constitutional ban upon unreasonable searches and seizures as a protection against invasion of privacy, and implicitly placed an equally strict quarantine upon the use in all courts, state and federal, of any evidence secured through any constitutionally prohibited intrusions.

In Frank v. Maryland,[61] the Supreme Court, after declaring an ordinance valid which authorized health inspections of dwellings, said:

"Two protections emerge from the broad constitutional proscription of official invasion. The first of these is the right to be secure from intrusion into personal privacy, the right to shut the door on officials of the state unless their entry is under proper authority of law. The second, and intimately related protection, is self-protection: the right to resist unauthorized entry which has as its design the securing of information to fortify the coercive power of the state against the individual, information which may be used to effect a further deprivation of life or liberty or property. Thus, *evidence of criminal action may not, save in very limited and closely confined situations, be seized without a judicially issued search warrant.* (Emphasis supplied.)

"But giving the fullest scope to this constitutional right to privacy, its protection cannot be here invoked. *The attempted inspection of appellant's home is merely to determine whether conditions exist which the Baltimore Health Code proscribes.* If they do appellant is notified to remedy the infringing conditions. *No evidence for criminal prosecution is sought to be seized.* Appellant is simply directed to do what he could have been ordered to do without any inspection, and what he cannot properly resist, namely, act in a manner consistent with the maintenance of minimum community standards of health and wellbeing, including his own. Appellant's resistance can only be based, not on admissible self-protection, but on a rarely voiced denial of any official justification for seeking to enter his home. The constitutional 'liberty' that is asserted is the absolute right to refuse consent to an inspection designed and pursued

---

[60] 267 U.S. 643 (1961).
[61] 359 U.S. 360, 79 S. Ct. 804 (1959).

solely for the protection of the community's health, even when the inspection is conducted with due regard for every convenience of time and place. (Emphasis supplied.)

"Moreover, the inspector has no power to force entry and did not attempt it. A fine is imposed for resistance, but officials are not authorized to break past the unwilling occupant.

"Thus, *not only does the inspection touch at most upon the periphery of the important interests safeguarded by the Fourteenth Amendment's protection against official intrusion, but it is hedged about with safeguards designed to make the least possible demand on the individual occupant, and to cause only the slightest restriction on his claims of privacy."* (Emphasis supplied.)

The court did not say an officer may enter the premises of another at any time, on any pretense, or over the refusal of the occupant to permit peaceable entry, without benefit of a warrant to arrest or to search.

The court simply said a health officer, without objections by an occupant, may enter a building or go upon property at reasonable times, on reasonable cause, without a warrant for the limited purpose of making an inspection to ascertain whether conditions are present which do not meet minimum standards and may be dangerous to the health, welfare and safety of the public.

In Ohio *ex. rel.* Eaton v. Price,[62] an ordinance was held constitutional by an equally divided Supreme Court which permitted housing inspectors to enter, examine, and survey dwellings at any reasonable hour; this law was valid because the inspection contemplated by it was not an "unreasonable search."

Notwithstanding the earlier Frank and Eaton cases just discussed, in 1967 the U. S. Supreme Court ruled in favor of the appellant where the right of entry without a warrant involved a housing code inspector.[63]

Inspection of Commercial Premises

Inspection of commercial premises involves ramifications differing from those associated with private dwellings. A case in point involved a warehouse which the fire department wanted to inspect under authority of a City of Seattle ordinance granting the fire chief the right "to enter all buildings and premises except the interior of dwellings as often as may be necessary." The owner refused to permit an inspection on the grounds that the ordinance was invalid. He said the fire chief had no search warrant, nor any probable cause to believe that a violation of any law existed on his premises. In upholding the constitutionality of the ordinance, the Washington Supreme Court (later overruled by the United States Supreme Court) said:[64]

"The purpose of the fire code inspection is to correct conditions hazardous to life and property. The problem of keeping cities and their inhabitants free from explosions and fires is a serious task facing all fire departments. It is obvious that routine inspections are necessary to insure the safeguarding of life and property. The need to conduct routine inspections of commercial

---

[62] Mem., 360 U.S. 246, 79 S. Ct. 978 (1959).
[63] Camara v. Municipal Court of the City and County of San Francisco, 387 U.S. 523, 87 S. Ct. 1727 (1967).
[64] See v. City of Seattle, 387 U.S. 541, 87 S. Ct. 1737 (1967).

premises, in regard to which probable cause for the issuance of a warrant could not ordinarily be established, outweighs the interest in privacy with respect to such premises. The purpose of the inspection contemplated by the code is not unreasonable."

The U. S. Supreme Court reversed the conviction of the warehouse owner on the ground that the Seattle ordinance, authorizing a warrantless inspection of his warehouse, was an unconstitutional violation of his rights under the Fourth and Fourteenth Amendments. The court cited its recent decision in Camara v. San Francisco[65] and declared that, "The businessman, like the occupant of a residence, has the constitutional right to go about his business free from unreasonable entries upon his private commercial property."

The Court felt that its recent decisions restricting administrative agencies in their attempts to subpoena corporate books and records supported their view that any agency's particular demand for access should be measured in terms of probable cause to issue a warrant, against a flexible standard of reasonableness that takes into account the public need for effective enforcement of the particular regulation involved. "But the decision to enter and inspect will not be the product of unreviewed discretion of the enforcement officer in the field."

"We therefore conclude that administrative entry, without consent, upon the portions of commercial premises which are not open to the public may only be compelled through prosecution or physical force within the framework of a warrant procedure. We do not in any way imply that business premises may not reasonably be inspected in many more situations than private homes, nor do we question such accepted regulatory techniques as licensing programs which require inspections prior to operating a business or marketing a product. Any constitutional challenge to such programs can only be resolved, as many have been in the past, on a case-by-case basis under the general Fourth Amendment standard of reasonableness. We hold only that the basic component of a reasonable search under the Fourth Amendment—that it not be enforced without a suitable warrant procedure—is applicable in this context, as in others, to business as well as to residential premises. Therefore, appellant may not be prosecuted for exercising his constitutional right to insist that the fire inspector obtain a warrant authorizing entry upon appellant's locked warehouse."

Inspection Program and Procedures

As a result of the above decisions, it might be advisable to establish a definite program for conducting inspections, and to set forth in a procedures manual the basis and frequency for making inspections for prefire planning, code enforcement, permit issuance, etc., whether by fire prevention bureau personnel in making technical surveys, or by fire fighters in making company inspections within their first-in districts. In the latter activity, it may be well to schedule prefire planning of hazardous occupancies at regular intervals, but only after explaining the need to the plant's management for obtaining a preview of con-

---

[65] See note 63.

ditions that will enable you to cope with future fires or explosions more effectively. Done properly, there should be no difficulty in getting an appointment for a guided tour for one or more companies by a company or battalion commander who is adept at good public relations.

With reference to the checking for fire hazards by fire station personnel, the procedures manual might call for inspection of target hazards, such as lumberyards, chemical warehouses, etc., four times a year; ordinary commercial occupancies biannually; and residences annually. The time interval is not so important as the fact that the inspections are scheduled on a regular basis and for a logical reason; should entry be refused and the need arise to go to a magistrate to obtain a warrant, there would be no necessity for citing in the warrant the violation of specific ordinances, since the probability that a hazard might exist could be indicated by the type of occupancy, past fire record of that or similar occupancies, the date of the last inspection, and the character of the area in which it is located. Excerpts from the procedures manual could be shown when applying for a warrant on those rare occasions when one might be required.

With the exception of such special occupancies as schools, hospitals, sanitariums, nursing homes, homes for the aged, boarding homes, etc., which require the issuance of a permit by the fire department, all routine building inspections could be programmed on the basis of an orderly geographical approach. If the inspections are planned so that the fire fighters will proceed through their districts on a building-to-building—block-to-block—basis, then when that rare objection is voiced on the part of some building owner to having his premises inspected, it can be easily shown that his building is not being "singled out" for a shake-down, but was reached as a part in the normal course of the planned sequence of inspections.

Right of Entry Ordinances

Following the Camara and See decisions, a number of cities have amended their ordinances relating to a fire inspector's right of entry, or to administrative inspections in general. For example, the Portland, Ore., Fire Code was amended in 1967 as follows:

(a) Any fire inspector, when engaged in fire prevention and inspection work, is authorized at any and all reasonable times to enter and examine any building, vessel, vehicle or place for the purpose of making inspections. *Except when an emergency exists, before entering a private building or apartment the inspector shall obtain the consent of an occupant thereof or a warrant of the Municipal Court authorizing his entry for the purpose of inspection. As used in this section, "emergency" means unforeseen circumstances which the inspector knows or has reason to believe exist, and which reasonably may constitute an immediate danger to public health or safety.*

(b) The official badge and uniform of the Bureau of Fire, when lawfully worn, shall authorize any inspector of the Bureau of Fire to enter and inspect buildings, vessels, vehicles and premises, as herein set forth.

(c) It is unlawful for any person to interfere with or attempt to prevent any inspector from entering and inspecting any building, vessel, vehicle or premises, *or, when an emergency exists or the inspector exhibits a warrant authorizing entry, from entering and inspecting any private building or apartment.*

(d) It is unlawful for any unauthorized person to wear the official badge or uniform of the Bureau of Fire, or to impersonate a fire inspector with the intent of unlawfully gaining access to any building, vessel, vehicle or premises in the city of Portland.

Administrative Inspection Warrants

Following the See and Camara decisions, a number of states have adopted statutes which spell out the allegations which have to be made by a public officer in an affidavit in order to obtain the issuance of an inspection warrant. In North Carolina, for example, the state legislature adopted a law which provides for the issuance of warrants for inspectors to conduct administrative inspections.

Before such a warrant can be issued, the issuing officer must be satisfied that the property to be searched or inspected is included in an authorized inspection program which naturally includes the property, or that there is probable cause for believing that there is a condition, object, activity or circumstance which legally justifies such a search or inspection. Moreover, the warrant must describe the property which is to be inspected with sufficient accuracy that the owner of the property can reasonably determine what property it has reference to, and must indicate the conditions, objects, activities, or circumstances which the inspection is intended to check or reveal; these warrants are valid for a period of twenty-four hours after issuance, and must be personaly served upon the property owner or possessor between the hours of 8:00 A.M. and 8:00 P.M., and must be returned within 48 hours.

Searches and Seizures

Frequently it is the duty of fire inspectors to apprehend violators of laws prohibiting the sale or possession of dangerous fireworks. It is well, then, for them to keep in mind the constitutional limitation against unreasonable searches and seizures, which in most states is identical with the Fourth Amendment to the United States Constitution.

Where the inspector has made a lawful arrest, with or without a warrant, he has the incidental right of search and seizure without a warrant, and this right extends to the person of the accused.[66] But a search of the house of the arrested party cannot be made without a search warrant[67] unless the arrest takes place on the premises.

Hence, if there is no lawful arrest and no permission given, the fire inspector should obtain a search warrant before beginning a search. In the case of a

---

[66] Weeks v. United States, 232 U.S. 392, 34 S. Ct. 341 (1913).
[67] Angello v. United States, 269 U.S. 20, 46 S. Ct. 4; Silversmith Lumber Co. v. United States, 251 U.S. 358, 40 S. Ct. 192 (1920); see 8 Cal. L. Rev. 347.

dwelling, mere suspicion and belief are not sufficient grounds for issuance of a warrant.[68]

In a case involving search of premises and arrest without a warrant, the U.S. Supreme Court said:

"The point of the Fourth Amendment, which often is not grasped by zealous officers, is not that it denies law enforcement the support of the usual inferences which reasonable men draw from evidence. Its protection consists in requiring that those inferences be drawn by a neutral and detached magistrate instead of being judged by the officer engaged in the often competitive enterprise of ferreting out crime. *Any assumption that evidence sufficient to support a magistrate's disinterested determination to issue a search warrant will justify the officers in making a search without a warrant would reduce the Amendment to a nullity and leave the people's homes secure only in the discretion of police officers.* Crime, even in the privacy of one's own quarters, is, of course, a grave concern to society, and the law allows such crime to be reached on proper showing. The right of officers to thrust themselves into a home is also a grave concern, not only to the individual but to a society which chooses to dwell in reasonable security and freedom from surveillance. *When the right of privacy must reasonably yield to the right of search is, as a rule, to be decided by a judicial officer, not by a policeman or Government enforcement agent."* (Emphasis supplied.)[69]

In 1971, a county district attorney in California was asked whether fire fighters could legally seize contraband found in plain view inside a private residence during the course of fighting a fire, and whether this would justify calling for police to search the remainder of the premises. C. Robert Jameson, district attorney for Yolo County answered as follows:

"When an emergency exists by fire or other hazard, and firemen or other emergency personnel enter a private residence in the performance of their official function, they may seize any contraband observed either in plain sight or subsequently discovered by them in the course of suppressing a fire, preventing reignition of a fire, determining its cause, preventing its spread, safeguarding other inflammable materials, discovering any dangerous condition, and searching for potential victims of the emergency.

"In such situations, and under these conditions, a fireman or emergency personnel may invite the entry of a policeman who may search the premises if his entry and search bears some reasonable relation to the purpose of dealing with the emergency. However, if the emergency is terminated, the fireman's prior lawful entry pursuant thereto will not justify police entry and activity without a search warrant.

"Firemen may properly report any suspicious condition to the police and if the emergency has terminated, a search warrant may be secured by the police for the search of the remainder of the residence, supported therefor by affidavit or oral testimony presented to the magistrate issuing the search warrant."[70]

---

[68] Nathason v. United States, 290 U.S. 41, 54 S. Ct. 11 (1933).
[69] Johnson v. United States, 333 U.S. 10, 68 S. Ct. 367 (1948).
[70] Jameson, Robert R., "Search and Seizure," *California Fireman,* July 1971, p. 5.

After pointing out that there is ample authority for policemen to seize such contraband as narcotics, machine guns, etc., Mr. Jameson declares that in those states where fire fighters are granted the powers of peace officers, such as provided in California for members of a fire department assigned to arson investigation and law enforcement duties, they would also have not only the authority, but the duty, to seize any contraband that is in plain sight.

Cases

An explosion and fire occurred and, fire fighters entered a residence under emergency conditions. In looking for potential victims, a fire fighter opened a wardrobe closet and discovered dynamite, machine guns, and ammunition. Police were summoned to gather the items and they were later admitted into evidence.[71]

Where fire fighters observe smoke or fire emanating from an otherwise constitutionally protected structure, they are clearly justified in entering pursuant to an apparent emergency.[72] In such situations the powers of fire fighters are those of both fire fighters and policemen.[73]

"Firemen enter under a license given by law for the benefit of the public generally and their right to enter is independent of any permission of the possesser of the property, who has no right to exclude them; that being lawfully in the building, it was the duty of the police and the firemen to protect the public, investigating to see that the fire would not start again and therefore to determine its cause, to check such conditions as gas and electrical installations, to safeguard inflammable materials, and to discover any dangerous conditions. A search for such purposes was lawful and was proper and the information discovered in making it was lawfully obtained."[74]

There is little doubt that evidence seized during a fire fighting operation, or during the later period of investigation, as long as fire fighters are on the premises, is likely to be admissible in court. The question is sometimes raised as to the authority of the police to continue the search after the fire fighters have gone back to quarters, unless, of course, a warrant is obtained for this purpose. Some courts are more liberal than others in extending the period of permissible investigation without a warrant. Here is an example: Following a residential fire in Newport, Ore., where the destruction was such that the defendant moved out of the rental unit, the fact that the landlord gave the fire and police officials permission to take photographs without obtaining a search warrant was not held to bar their use in the arson trial. The court ruled the defendant had actually abandoned tenancy the day of the fire and his consent was not required. Although the officials visited the scene over a period of thirty-six days, the court said that since the first visit of the fire chief, where evidence of arson was very apparent, the later visits were merely a continuation of the first inspection and did not amount to an unconstitutional search.[75]

---

[71] Romero v. Superior Court, 266 Cal. App.2d 714, 72 *Cal. Rptr.* 430 (1968).
[72] People v. Ramsey, 272 Cal. App.2d 302 (1969).
[73] See note 71, at p. 722.
[74] See note 71 at p. 719, citing State v. Cohn, S.W.2d 691 (Missouri 1961).
[75] State of Oregon v. Felger, 526 P.2d 611 (Oregon App. 1974); Fire Dept. Pers. Rep.,Jan. 1975, p. 10.

There is substantial agreement that when a fire fighter investigates a fire, whether as officer in charge at the scene, or as a representative from the arson bureau, fire marshal's office, or other capacity, he can seize evidence relating to a possible crime that is plainly visible without a warrant. This does not mean, though, that he can search all the drawers and cupboards in a dwelling on the chance that he might find evidence of a crime. But if what he finds in the open leads him to believe that a crime has been commited, he can obtain a warrant to search the rest of the premises more thoroughly. Sometimes the evidence picked up off the floor will suffice. For example, there is a Delaware case where the defendant objected to the submission of evidence that was obtained without a warrant, but to no avail. The state fire marshal picked up broken gasoline bottles at the fire scene during and immediately after the fire, and their introduction as evidence in the trial resulted in the conviction of the defendant for murder of his daughter-in-law. The Third Circuit Court in 1974 said that the search and seizure was justified by the exigent circumstances, and denied a writ of habeas corpus for his release.[76]

Right to Photograph

There is no question that fire investigators may take photographs during and following extinguishing operations as an incidental tool in establishing the cause and origin of the fire, or to be used in evidence in establishing the commission of a crime; however, where a fire has not yet occurred, and the only purpose is to enter the premises to photograph fire hazards, some objections may be raised by the occupant. In the absence of specific provisions in the ordinance or statute granting fire fighters the right to enter, examine, inspect, and photograph premises for fire hazards, it is doubtful that they would have the right to take pictures. Therefore, it would be unwise to arrest an occupant for refusal to permit the taking of pictures on his premises unless specific statutory authorization exists for this procedure.

Entrapment

Since it has been held not to be entrapment for plain-clothes men to place a bet with an illegal gambling house, nor to write for obscene pictures of which the defendant is suspected of sending through the mail, and use the pictures to convict him, it would not seem illegal for fire inspectors to induce the sale of firecrackers in violation of ordinances or a state law by a person suspected or reasonably believed to be engaged in such illegal sales.

The court allowed the officers in the foregoing cases to encourage the accused in a course of conduct already begun on his own volition; it does not allow them to start the accused on a course of criminal conduct and then penalize him for it. It is well settled, however, that where the officers merely induce the defendant to sell the outlawed article, entrapment is no defense;[77] where, however, a prohibition agent repeatedly entreated an old war "buddy"

---

[76] Steigler v. Anderson, Warden, 496 F.2d 793 (3rd Cir. 1974).
[77] People v. Ramirez, 95 Cal. App. 140, 272 P. 608 (1928).

to buy him a gallon of whiskey, and then arrested him for it, the court condemned such tactics as an abuse of authority.[78]

Investigation Authority

Statutes and ordinances authorizing civil inspections have long been acknowledged and sanctioned as incident to the police power of a state or municipality, but they must be within constitutional limits. The Supreme Court of Iowa (in a five to four decision)[79] refused to suppress evidence obtained by fire investigators which was claimed to have been gained through unlawful search and seizure over a period of five or six weeks.

After discussing several cases on which the constitutionality of laws granting civil authorities the right to make inspections without a search warrant were based, the court held that the statutory authority to make an investigation does not end when the fire is extinguished and evidence of arson was found, since the original entry was lawful and reasonable; the fact that the investigation becomes accusatory does not make it unconstitutional. The court went on to declare:

"It does go beyond proper limits when it extends into fields unrelated to the authorized investigation or is unduly prolonged over the objection of the accused. Not even a search warrant is good indefinitely. The question is not before us but it may be that the defendant could have, after a reasonable time for completion, terminated the investigation by telling the investigators to stay out. In that event further search and seizure would have required a search warrant. However, the record is silent as to any search beyond the limits of the statutory mandate or over any objection, timely or otherwise, by the defendant."

In commenting on the case of Mapp. v. Ohio,[80] the Iowa court noted that the United States Supreme Court held that evidence obtained by unconstitutional search is inadmissible, and that the rights guaranteed by Amendment 4 are enforceable against the states by the Due Process Clause of Amendment 14. Then it added:

"With these basic rights of the people no one has any legitimate quarrel. However, neither the Constitution nor the cases decided thereunder say that there can never be a legal or reasonable search without a warrant. Neither do they say that evidence obtained during a legal investigation is inadmissible. The constitutional guarantees do protect against indiscriminate and unreasonable searches and against warrants issued without probable cause."

Thus, it is the indiscriminate search and seizure without benefit of a warrant issued on affidavit of probable cause that is prohibited. There is no rule against the use of evidence obtained during a reasonable and legally authorized investigation. The Rees case contains this observation: "It is unreasonable and illogical to say that when officers are carrying on a legal or as in this case a mandatory investigation they must stop and get a search warrant before they

---

[78] Sorrells v. United States, 57 F.2d 973, 287 U.S. 435, 53 C. Ct. 210, 86 A.L.R. 249 (1932).
[79] State v. Rees, 258 Iowa 813, 139 N.W.2d 406 (1966).
[80] Mapp. v. Ohio, 267 U.S. 643 (1961).

can seize and later use the evidence for which they were making their original investigation.

"The test is the reasonableness of the search under proper authority and not the source of the authority under which the search is made. A reasonable search mandatory under a legislative enactment is clothed with as much dignity and is entitled to as much consideration as a search under a warrant issued by a Justice of the Peace. It would be illogical to say that the evidence would have been admissible if seized under a search warrant but inadmissible if seized under the statute. This assumes, of course, that the entry was reasonable and legal and no one contends otherwise. Within the constitutional limitations as to reasonableness the legislature may and has authorized such an investigation as was made here."[79]

Interrogation Limits and Rights of Suspects

Although the recommended warnings which should be given to a person accused of criminal activity following his arrest, and prior to interrogation, are presented in Chapter 14, it might be well at this point to summarize the decisions which have been handed down by the U. S. Supreme Court in relation to the rights of suspects in criminal cases. After the brutal treatment which was given the defendants in Brown v. Mississippi,[81] for the purpose of getting a "voluntary" confession, the court held that they had been denied due process of law. But since the state courts continued to admit confessions obtained by third degree methods, the Supreme Court extended the Brown decision from admitted physical brutality to cases involving admitted prolonged "grilling," intimidation, etc.

In 1943, the Court was confronted with a case where the defendants were subjected to prolonged detention and interrogation resulting in a confession; so in McNabb v. United States[82] it ruled that prolonged detention violated a federal statute which required arrested persons be promptly taken before a magistrate. In spite of this, some of the lower federal courts refused to hold that a confession was inadmissible solely on the ground that it had been made during a prolonged detention, so the Supreme Court in 1957 declared in Mallory v. United States[83] that any confession made during a period of unnecessary delay in taking the defendant before a magistrate was inadmissible in a federal criminal trial.

Since the McNabb-Mallory rule did not apply to state criminal cases, it took Mapp v. Ohio, discussed in the preceding section, to exclude evidence obtained by unconstitutional methods. In the case of Gideon v. Wainwright,[84] the Court held that the Sixth Amendment right to counsel was applicable to the states; in Escobedo v. Illinois[85] it ruled that the suspect's right to a lawyer applied in the police station, and that evidence obtained in violation of this right must be excluded from evidence.

---

[81] 279 U.S. 278 (1936).
[82] 318 U.S. 332 (1943).
[83] 354 U.S. 449 (1957).
[84] 372 U.S. 335 (1936).
[85] 378 U.S. 478 (1964).

In 1966 in Miranda v. Arizona[86] the Court held that the right to counsel in the station house meant that the suspect must be warned that he could have a lawyer, and that he would be entitled to have one appointed for him if he could not afford to hire his own. It also laid down strict standards governing the waiver of these rights by the accused. Also in 1966 the Court applied the Miranda rule to juveniles, declaring that police interrogation must stop when the suspect asks to see one of his parents, and in 1972 extended the right to counsel to misdemeanor as well as felony cases.

As a result of this trend toward increasing the rights of the defendant for the purpose of bettering his chances of obtaining a fair trial, there developed some opposition on the part of "law and order" oriented persons, including some judges. An indication of this feeling is expressed in the following excerpt from the dissenting opinion of Judge Warren E. Burger (now chief justice of the U. S. Supreme Court), in the case of Frazier v. United States:[87]

"The seeming anxiety of judges to protect every accused person from every consequence of his voluntary utterances is giving rise to myriad rules, sub-rules variations and exceptions which even the most alert and sophisticated lawyers and judges are taxed to follow. Each time judges add nuances to these rules we make it less likely that any police officer will be able to follow the guidelines we lay down. We are approaching the predicament of the centipede on the flypaper—each time one leg is placed to give support for relief of a leg already "stuck," another becomes captive and soon all are securely immobilized. Like the hapless centipede on the flypaper, our efforts to extricate ourselves from this self-imposed dilemma will, if we keep it up, soon have all of us immobilized. We are well on our way to forbidding any utterance of an accused to be used against him unless it is made in open court. Guilt or innocence becomes irrelevant in the criminal trial as we flounder in a morass of artificial rules poorly conceived and often impossible of application."

Since 1971, the Supreme Court has appeared to retreat from the controversial Miranda guidelines. In the first departure, the Court held in a five to four ruling, that illegally obtained statements which are inadmissible as evidence against a defendant because of a failure to warn him of his right to remain silent and to counsel, can be used to attack his credibility at the trial. The opinion, written by Chief Justice Burger, limited the effect of the Miranda decision, and included the following statement: "The shield provided by Miranda cannot be perverted into a license to use perjury by way of defense, free of the risk of confrontation with prior inconsistent utterances."[88]

In 1974 the Court held that the Miranda decision did not require the suppression of testimony by a witness whose name was learned by police during the illegal questioning of the suspect.[89] In the early part of 1975 the Court allowed prosecutors to introduce during the trial of a suspected bicycle thief a confession obtained from him after he had demanded to see an attorney.[90] The

---

[86] 384 U.S. 436 (1966).
[87] 419 F.2d 1161 (D. C. Cir. 1969).
[88] Harris v. New York, 401 U.S. 222, 91 S. Ct. 643 (1971).
[89] Michigan v. Tucker, —— U.S. —— 94 S. Ct. 2357 (1974).
[90] Oregon v. Hass, —— U.S. —— 95 S. Ct. 1215, (1975).

evidence was used only to impeach the credibility of the defendant on the witness stand when he blamed the theft on his friends.

On December 9, 1975, the Supreme Court again narrowed the Miranda guidelines, with two dissenters predicting its ultimate demise, in ruling that it was permissible for the police to resume interrogation after the suspect had exercised his constitutional right to remain silent under the Fifth Amendment. In writing the opinion for the Court, Justice Potter Stewart declared that the privilege against self-incrimination is not violated as long as police officers "scrupulously honor" the suspect's renewed protests, should there by any, that he wanted the questioning to stop. In this case, two hours after the suspect had invoked his privilege to remain silent, the police told him, inaccurately, that an accomplice had named him as the gunman in the murder case, and under further questioning, after reminding him of his rights, he made several incriminating statements. The dissent argued that once a suspect has insisted on silence, no further questioning should be allowed until an attorney is summoned to advise him, and expressed the fear that the majority opinion encourages police to continue the suspect's detention until the police station's "coercive atmosphere" does its work and the suspect collapses.[91]

In view of the above decisions, and notwithstanding the apparent erosion of the guidelines laid down in the earlier cases, it would be well for the fire department's law enforcement officers to follow the suggestions set forth in Chapter 14 of this book with respect to the interrogation of persons who are suspected of any criminal activity, whether felony or misdemeanor.

Failure to Enforce Ordinances

Neither the city nor the fire inspector is liable in damages for a failure to enforce ordinances which have been enacted by the city. In one case a court held that there was no liability for injury to one struck by a skyrocket due to the failure to enforce an ordinance forbidding the setting off of fireworks.[92] But a city was held liable where a little girl was struck in the face by a skyrocket discharged in the streets by a labor union which had obtained a permit from the city to shoot the fireworks in a vacant lot. Since the city should have anticipated a vast crowd at the display, it had granted a use foreign to the use of the street, thereby failing to keep it in a safe condition, and in effect created an actionable nuisance.

In other cases it has been held that there is no liability for such things as failure to enforce ordinances relating to the storage of flammable liquids or the proper maintenance of skylights, nor for the improper issuance of a license because of a failure to follow the ordinance requirement that a bond be put up, nor for the failure to make an inspection and give a notice to the owner of a dangerous overhanging porch to have it repaired. In the latter case, the fire

---

[91] Michigan v. Mosley, —— U.S. ——, 96 S. Ct. 321 (1975).
[92] Gilchrist v. Charleston, 115 S.C. 367, 105 S.E. 741 (1920).

marshal's failure to inspect buildings in the city carefully resulted in serious injury to a fire fighter who climbed upon the defective porch and fell to the ground when it collapsed. The court said that the ordinance only required the fire marshal to give a notice or to abate the dangerous condition when he had knowledge of it. It was not shown that the condition was obvious nor that the fire marshal had knowledge of it. Since the same was true of the owner of the building, the fire fighter was unable to recover from either.

The above opinion would seem to indicate that, in some states at least, the fire inspector should be very careful to follow whatever duty is enjoined upon him by law with respect to giving the occupant or owner notice of any violation he finds upon the premises. It might not be considered sufficient to merely mention the violation to the owner if the ordinance requires that a written notice be served upon him, or a written report submitted to the inspector's superior officer. This matter becomes extremely important in case a disaster occurs, such as the Cocoanut Grove night club fire in Boston where the fire inspector was later indicted by a grand jury for failure to report violations of the building laws and to report insufficient exits.

In a Mississippi case a building inspector required the plaintiff to pour concrete in an improper way, resulting in great loss to the plaintiff. While the court held that the city was not liable for the tortious acts of building or fire inspectors, it did not say that an inspector was immune from liability for his wrongful conduct.[93]

In Maryland, however, even a municipality is liable for neglect to enforce ordinances passed in the exercise of its corporate powers, except where the power to enforce them is given to an independent board or officer.[94]

The Village of Liberty (New York) and its building inspector have been sued by the survivors of a residential fire for failure to enforce the ordinances relating to fire escapes, and the Appellate Court refused to dismiss the complaint on the grounds that the statutes and ordinances relating to fire safety were enacted for the special benefit and protection of those in the plaintiff's class, and the allegation that the Village's "known incompetent agent, the village building inspector, had knowledge of the conditions which constituted a nuisance and a fire trap and failed to take steps to remedy them." (Sanchez v. Village of Liberty, 375 N.Y.S.2d 901, Dec. 11, 1975.)

Duty of Public toward Fire Inspector

Does a fire inspector who goes upon the premises of another in the course of his duty and is injured by reason of some dangerous or defective condition, unknown to him but known to the owner or occupant, have a cause of action against the latter? The fundamental question is whether he is an invitee or a bare licensee. A person on the premises by invitation, express or implied, is an

---
[93] Bradley v. Jackson, 153 Miss. 136, 194 Fed. 775 (Md. 1911).
[94] State v. Miller, 194 Fed. 775 (Md. 1911).

invitee, and the owner owes him a duty of reasonable care; whereas one who is there by mere toleration or permission is a mere licensee, and the owner owes him no greater duty than to avoid wilful or wanton injury.[95]

While fire fighters are generally held to be licensees, "An inspector or official whose presence is necessary to the lawful conduct to the defendant's business is commonly held to be a business visitor or invitee to whom a duty of reasonable care to prevent injuries from defective conditions of the premises is owed."[96] This is on the basis that his presence is for the mutual benefit and advantage of the parties.

In an Illinois case, the owner of a building was liable for the death of a water inspector who fell on a belt in the engine room and was carried into a fly-wheel. In holding the inspector an invitee, the court distinguished an earlier case by asserting that the deceased did not enter the premises under a license or authority given by law, but was there for the common advantage of both parties.[97]

Is not the work of the fire inspector of mutual advantage to the city and the owner of the premises checked by him? Certainly in a city where regular inspections are made at fairly definite intervals the landowner can anticipate the inspector's presence, thus removing the objection commonly urged against classifying a fire fighter as an invitee.

A California case[98] involving a boat owner, while avoiding the issue of whether a customs inspector was an invitee or licensee, nevertheless imposed liability on the ground that his presence was known to the boat owner, and therefore there was a duty to provide a safe gang-plank. The court said that where a man goes on the premises, though with an absolute right, but fails to notify the owner, he comes under the rule applicable to trespassers or mere licensees. Therefore it is important that the fire inspector, before proceeding with his tour of the premises, go to the occupant and make his presence known. This provides the opportunity to be warned of hidden dangers, and permits him to take such measures as are possible to prevent injury.

Meat inspectors, as well as federal revenue inspectors at a liquor distillery, have been held to be invitees, and so it would seem that the tendency is to increase the landowner's liability toward government inspectors.

However, an inspector who enters a dark building that is in the course of construction, groping through a passageway into which he has never before walked, and which from previous observation he knew was devoid of flooring, is guilty of contributory negligence as a matter of law, and is barred from a recovery if he falls through to the basement below.[99]

It would seem incumbent on the inspector to carry a flashlight, or else refrain from entering into dark places, if he would avoid losing his right of recovery based on the landowner's negligence.

---

[95] Litch v. White, 160 Cal. 497, 117 P. 515 (1911).
[96] "Inspectors—Licensees or Invitees?," 22 Min. L. Rev. 88; see also "Restatement of Torts," Sec. 345, where fire fighters are given status of licensees.
[97] Kennedy v. Heisen, 182 Ill. App. 200 (1913).
[98] Wilson v. Union Iron Works Drydock, 167 Cal. 539, 140 P. 250 (1914).
[99] Ross v. Becklenberg, 209 Ill. App. 144 (1917).

Review Activities

1. Write a brief comparison of "common law liability" and "absolute liability." Illustrate your description by citing examples of each.
2. In a general class discussion, elicit answers to each of the following:
   (a) What is one of the primary sources for an effective fire prevention program?
   (b) What are some of the means of proving a building is a public nuisance?
   (c) Name several buildings in your community that constitute a public nuisance. Explain why.
   (d) How would you go about proving that the buildings named in (c) are public nuisances?
3. (a) Through the exercise of its police power, what areas does a municipal fire department have charter rights to administer ordinances on?
   (b) What areas of fire prevention can a fire department enforce?
4. Research the latest federal decisions concerning "Right of Entry" by members of the fire services. Summarize your findings in outline form. For purposes of comparison and discussion, circulate your outline within a small group of your classmates.
5. (a) What rights does a fire inspector have in inspecting private dwellings? Commercial dwellings?
   (b) What is the proper and legal way for a fire inspector to conduct an inspection?
   (c) When is the owner or an occupant of a dwelling liable if a building inspector is injured or killed on the premises?
6. Distinguish between federal, state, and local regulation in the field of fire prevention. Cite an example of jurisdictional conflict other than the one mentioned in the text.

Chapter Nine
PENSIONS

Pension and Compensation Distinguished

Notwithstanding the fact that fire fighters' retirement pensions have sometimes been referred to as "delayed compensation," the words "pension" and "compensation" are not synonymous, and therefore an award for injuries under a workmen's compensation act will not preclude the recipient from becoming a beneficiary of a pension plan.

If, however, the combined pension and compensation award is in excess of the amount permitted by law, then the pension board may recover the amount overpaid by deducting it from future installments by decreasing amounts paid minimally, and by using this method over a period of time. And if a city charter provision bars a dependent from a pension where the injury which caused the death was not compensable under the workmen's compensation law, the dependent has no option to take one or the other.

In a 1965 California case, the city of Los Angeles claimed that it did not have to pay workmen's compensation claims so long as it paid pension benefits. The local union contested this stand and won. The Supreme Court of California held that the city would be liable for workmen's compensation claims plus that portion of the pension which had been contributed by the individual involved. In Los Angeles, the fire fighters and police paid 6 percent of their salary into the pension fund and the city paid an amount equal to 13 percent of the salary. The court ruled that the city would have to pay that amount proportionate to the employee's contribution.

Constitutionality of Pension Plans

Retirement and disability pensions for fire fighters are generally governed by statute or charter provisions, and are sustained against attacks on their constitutionality, insofar as they are held a lawful appropriation of municipal funds, not only as to the original pension grant, but as to subsequent increases in amount. Statutes requiring cities to pension fire fighters have also been held constitutional. A 1949 Pennsylvania case has held that compulsory retirement of fire fighters under civil service at the age of sixty or after twenty years' service can only be accomplished by a general, nondiscriminatory ordinance which operates uniformly and equally upon all the members of the class created by the ordinance.

By interpreting a state law which provided that no city of the Second Class could remove a policeman or fire fighter without his written consent or a court order as being applicable to twelve fire fighters who were put on retirement against their will because of insufficient funds, the Commonwealth Court of Pennsylvania held that their compulsory retirement was illegal and ordered the City of Scranton to reinstate them to the fire bureau with back pay from the date of the dismissal.[1]

Pension Provisions Liberally Construed

Since pensions are a matter of statute and not of right, one who seeks to be a beneficiary under a given law must establish his compliance with all the conditions relating to such matters as term of service, disability in line of duty, proper and timely presentation of claim, etc. Although "the purpose of pension laws is beneficial, and statutes of this character should be liberally construed in favor of those intended to be benefited," they are not deemed to take effect retroactively so as to include fire fighters who have left the service at the time of the enactment.[2] They are, however, interpreted so that, with reference to the fire fighter in the department at the time the pension law is adopted, the required service period dates from the beginning of the fire fighter's employment.

Personnel Entitled to Pension

A question that has often arisen with respect to a claimant's right to a pension is whether he is a "member" of the fire force within the meaning of the law. Through liberal construction of fire department pension provisions, the following have been held to be entitled to its benefits: blacksmith helper, extra man, janitress, veterinary, lineman, watchman and veterinary surgeon, fire alarm crew, harness makers, and a clerk or secretary. While it may thus be seen that actual fire fighting is not always the criterion applied in determining whether a fire department employee is entitled to share in the pension benefits, yet the California Appellate Court, for example, expressed its reluctance to extend such benefits to San Francisco engineers and pilots of fireboats in the following language: "Relief and pension systems for firemen are to be liberally construed for the benefit of firemen, but it is of no advantage, and is in fact a disadvantage, to have included within the pension system a number of men who are not exposed to danger."[3] And in a Massachusetts case, members of the repair division of the Boston Fire Department were held ineligible to a retirement pension on the ground that "members" of the fire department, within the meaning of the statute, included only those fire fighters who wear a uniform, attend drill school, take a physical examination before appointment, and who are not entitled to workmen's compensation.[4] The court noted that

---

[1] Bauer v. Peters, 331 A.2d 245 (1975).
[2] *In re* Anthony, 71 Utah 501, 267 P. 789 (1928); Schieffelin v. Enright (1921) 190 N.Y.S. 328, mod. 200 App. Div. 312, 192 N.Y.S. 729 (1922); Clarke v. Police Life and Health Insur. Bd., 127 Cal. 550, 59 P. 994 (1900).
[3] Carrich v. Sherman, 105 Cal. App. 546, 228 P. 143 (1930).
[4] Elliott v. Boston, 245 Mass. 330, 139 N.E. 523 (1923).

the essential difference in the work of the repair men and that of the fire fighters lay in the fact that the fire fighter's work is hazardous, and distinguished another case on the basis that the plaintiff in that case shared in a charitable fund and not one raised by taxes. A New Jersey case held that men in the public safety department who, after a governmental reorganization, were still performing the same fire fighting duties as they had done in the fire department, were nevertheless entitled to fire fighter's pensions.[5]

The Supreme Court of Pennsylvania, in holding that a fire inspector was a "regularly appointed member" of the City of Washington, Pennsylvania, Fire Department, and hence entitled to a pension, rejected the limited interpretation put on the Black's Law Dictionary definition of a "fireman" by the Fireman's Relief Association; this definition says that a fireman is a person engaged in the fighting and extinguishment of fires, and in holding this to be too limited, the court said: "How this definition leads the Association to conclude that 'fighting fires' is limited to fighting them by pulling hose, squirting water, or driving a truck, is hard to fathom. Fire fighting has become more and more complicated in the modern technological world."[6]

The test which should be applied in the above cases is not the origin of the fund nor the combat of flames, but the nature of the employee's duty, i.e., does he directly (not remotely) carry out a function assigned to the fire department by ordinance or charter provision? It could not be successfully contended that in a large municipal fire department, directed by law to not only extinguish fires, but prevent them and enforce fire ordinances, that only those men who come in contact with flames are "members" of the fire department. It requires no liberal construction to concede that fire inspectors, arson investigators, rescue squads, salvage crews, fire apparatus drivers, pump operators, and possibly alarm board men, all of whom directly engage in activities ordinarily required by charter provision, are intended to be included within the terms of the ordinary pension provisions. Yet the above groups of men do not necessarily come in direct contact with flames. On the other hand, it might exceed even the bounds of liberal construction to contend that the ordinary pension provisions for fire fighters are meant to include the vast stenographic force employed in the larger fire departments, or possibly the carpenters, painters, mechanics, and others who are only remotely assisting in executing the principal duties of fire fighters.

Although the chief of a fire department, who was also the city electrician, paid an amount into the pension fund based upon his combined salaries, he was not entitled to a dual pension, but allowed a refund of the excess payments.[7] Where, however, the position of superintendent of fire alarm system was abolished and the position of city electrician was created with the assumption of the duties of the former, the retired fire alarm superintendent was not entitled to have his pension increased to an amount proportionate to the salary of the new position.[8]

---

[5] Taggart v. City of Asbury Park, 15 N.J. Misc. 10, 188 A. 490 (1936).
[6] Morse, H. Newcomb "Legal Insight: Pension Rights," *Fire Command!*, Vol. 41, No. 7, July 1974, p. 32.
[7] State v. Welch, 181 Wis. 147, 194 N.W. 382 (1932).
[8] Watkins v. Pension Board of Sacramento, 91 Cal. App. 542, 267 P. 323 (1928).

Death or Injury in Line of Duty

In order to establish a right to disability pensions, or to those granted widows and dependents of fire fighters killed in action, it is usually required that the claimant prove that the injury or death occurred in line of duty, or as the result of an incident happening within the scope of his employment. But the question of what activities should be considered "in the performance of duty" is a matter of judicial controversy, and can only be settled in any given instance by an examination of the local decisions interpreting provisions of this nature.

Pneumonia, Heart Attacks, Cancer

Fourteen states have laws creating a presumption that certain injuries arise in the line of duty, such as hernia, pneumonia, heart attacks, etc.[9] Although a North Carolina court held that its state law violated a state constitutional provision that forbade special privileges for certain public service employees,[10] the Minnesota law was given judicial approval in 1941,[11] and the Maryland Supreme Court affirmed the constitutionality of its law in 1975.[12] The fact that the Maryland law was only applicable to fire fighters was held not a violation of the equal protection clause of the U.S. Constitution because fire fighters are exposed to health hazards not shared by other government personnel.

In the absence of a statute creating a presumption that pneumonia and heart attacks are the result of conditions arising in line of duty, and even where there are such statutes, if as in California, there is a marked tendency to rebut the presumption, it is often difficult to establish a casual connection between a specific incident occurring in the performance of duty and the resultant injury or death.

The term "injuries" has been held broad enough to include pneumonia contracted by fire fighters as a result of becoming overheated and chilled in the course of making building inspections. But where a period of two years had elapsed between the date of injury and the fire fighter's death from the grippe, or where the deceased person had in his lifetime neglected to make any record of the injuries which are later asserted to have caused his death from pneumonia, a finding by the board or commission that the deceased did not die of injuries received in line of duty will not be disturbed by the courts.

Although cases may be found to the contrary, yet, in view of the fire fighter's strenuous exertion on a moment's notice, coupled with the startling effect of being aroused into action by the sound of the fire gong, it can be readily understood why courts sustain the contention that death from heart trouble is a line of duty injury for fire fighters. Whatever predisposing physical condition may exist, if the employment is the immediate occasion of the injury, it arises out of the employment because it develops it.

Where the evidence showed that the disability was caused by a hypertrophic attack of arthritis of the spine, and attributed to infection of the teeth or tonsils,

---
[9] FIRE DEPT. PERS. REP., No. 7, July 1975, p. 6.
[10] Duncan v. City of Charlotte, 66 S.E.2d 22 (N.C. 1951).
[11] Kellerman v. City of St. Paul, 1 N.W.2d 378, 380 (Minn. 1941).
[12] Board of County Commissioners v. Colgan, 334 A.2d 89 (Md. 1975).

a fire fighter was held not entitled to a disability pension, although he contended that he had once been exposed to smallpox, twice to influenza, and suffered a foot infection—all in line of duty. In this case the court declared that "'disabled in the service of the city' is properly understood to mean 'disabled by reason of bodily injuries received in, or by reason of sickness caused by the discharge of the duties of such person.'"[13]

Likewise, where a St. Louis fire fighter claimed that his disability from a spinal injury was the result of a fall while on duty, but felt no pain until about five months later, the ruling by three doctors that the injury was not work related was upheld by the Missouri Appellate Court, notwithstanding the plaintiff's objection that the medical records were not formally presented to him for inspection before submitting them into evidence. At an administrative hearing it is not necessary to observe all the formalities that are customary in court. The important question is whether the reports constituted substantial evidence upon which the Board could deny the pension, and the court found the decision reached was not against the overwhelming weight of the evidence.[14]

In another case, also involving the medical panel of the pension board of the St. Louis Fire Department, a fire fighter claimed that his heart disease was caused by a fall while on duty almost two years before he complained to his physician of discomfort. The Missouri Appellate Court again affirmed the circuit court's decision to uphold the pension board's finding that the fire fighter's heart disease and spinal osteoarthritis were not caused by the fall. One doctor had attributed the heart disease to several possibilities including the assertion that he was carrying too big a load of fat.[15]

A California case held that cancer of the lungs, presumably caused by prolonged inhalation of smoke and gases at fires, was a line of duty disease, and that the wife of a deceased fire captain was entitled to established benefits.

Contagious Disease and Insanity

Considering the numerous possibilities of a fire fighter contracting a contagious disease, not merely in the course of extinguishing fires but also in rescue activities where calls are frequently received to resuscitate victims in their homes and in hospitals, the courts have justifiably held that a disease, such as syphilis, may have been contracted in line of duty so as to entitle a fire fighter suffering from insanity, as a result thereof, to a disability pension.

Suicide

Where a policeman was found dead at home with a shotgun between his legs and his head half blown off, his death was held to be "in the service of the department" within the ordinance for pensioning dependents, for there was no finding of suicide. But even suicide, where it can be shown that it was caused by an unbroken chain of events occurring in line of duty, may be the basis for

---
[13] Tripp v. Board of Fire Commissioners, 94 Cal. App. 720, 271 P. 795 (1928).
[14] Morse, H. Newcomb, "Legal Insight: Presentation of Records," *Fire Command!*, Vol. 39, No. 5, May 1972, p. 34.
[15] Morse, H. Newcomb, "Legal Insight: Heart Ailment Not Service-caused," *Fire Command!*, Vol. 39, No. 10, Oct. 1972, p. 30.

a pension claim on the part of a fire fighter's widow.

Thus in Baker v. Fire Pension Fund Commissioners,[16] where a fire fighter's back was broken when he was riding in a fire truck that overturned, with the back injury resulting in such pain that he committed suicide, he was considered killed in line of duty within the meaning of the pension for widows. The California court said about this particular case: "The injuries which Baker received may justly be said to have been the proximate cause of his death. They set in motion a train of events, operating from cause to effect, that, without the intervention of any outside independent cause, resulted in his death."

At Home

Where fire fighters are subject to call at all times, their situation may be deemed analogous to that of policemen who are allowed specified intervals to return to their homes. In an Ohio case there is precedent for holding that death occurring while at home on rest hours is in line of duty, even though the deceased person may have been in civilian clothes and justifiably killed while engaged in the act of committing a felony.[17]

On the Street

Like policemen, fire fighters also have bands, and it appears that if a band player is killed by an automobile while running across the street (to order some band music), his demise is in the performance of duty and his widow may obtain a pension. But if he is killed in an auto accident on the way home to dinner, or falls off a street car during the brief period out of the twenty-four hours in which he is allowed to go home for his meals, then death resulting from the fatal injury has been held to be outside the performance of his duty.

In an Ultra Vires Activity

Likewise, where a fire fighter was killed while riding with his chief to obtain some grappling hooks with which to retrieve some drowned bodies, his death was said to have occurred while acting beyond the scope of his employment, the latter having been limited to protecting lives and property from fire.

Delay in Presentation of Claim

Failure to make a timely presentation of a claim before the proper officer or commission may bar the claimant's remedy to enforce his right to a pension. It is essential to allege and prove that a demand for a pension has been presented and rejected in accordance with the charter requirements.

The Los Angeles city charter provisions were construed to bar any suit upon a claim not commenced within three years (excluding the time taken by the board to deliberate on the application) after the cause of action accrued, the court saying,[18] "the cause of action to establish the right to a pension

---
[16] 18 Cal. App. 433, 123 P. 344 (1912).
[17] Hefferman v. State, 38 Ohio App. 552, 177 N.E. 43 (1931).
[18] Dillon v. Board of Pension Commrs. of Los Angeles, 5 Cal. Dec. 281, 116 P.2d 37 (1941).

accrued to the plaintiff at the time of her husband's death." More than three years had elapsed since the claim had been presented, but the action was filed within three years of the claim's rejection. The court declined to give proper consideration to the continuing nature of pension claims and failed to adequately distinguish a previous California decision, known as the Dryden case,[19] which held that a claim may be filed at any time for installments falling due within the last preceding six months, and another California decision, known as the Talbot case,[20] holding that the three-year statutory period did not render a claim demurrable with respect to recent accruals. In the latter case, the doctrine of laches was held inapplicable to recent accruals.

The liberal view was followed in an Iowa case, where a court held that a widow's right to a pension is not subject to any time limit, and that it would continue to exist all during her life. In a similar case involving the widow of a policeman, a delay of two years prior to presenting the claim was held to be no bar under the statute of limitations, nor by laches.

But in a Minnesota case, where a period of twelve years had elapsed since the claim had been filed, the continuing nature of a pension was not considered, though the court declared that a cause of action of this nature arises at the time of the refusal to put the claimant on the pension roll.[21] Similarly, periods of three years, nine years, and nineteen years, were held to be excessive delay so as to bar the claims for pensions. One court applied the analogy of easements, and held that a failure to apply for such a pension for seven years resulted in the claimant's losing his right by abandonment.

Termination of Employment as Affecting Pension Rights

Where the applicable provisions require that a person be a member of the department at the time of application for a pension, dismissal from the service prior to presentation of the claim, whether the required period of time had been served, or not, will act as a bar in some jurisdictions, as will, likewise, the abolition of his position.

Because an Oklahoma fire fighter had turned in his equipment about three years before the required period of service had elapsed, and had not had his name certified for re-appointment by the chief, he was deemed removed and pension denied, notwithstanding the fact that he had continued to respond to fires during those three years.[22]

Voluntary resignation, which, in the absence of a specified date, is said to take effect when one copy of the resignation is deposited in the office of the fire department and another in the office of the civil service commission, ends all right to a pension in some cities. A subsequent withdrawal of the copy of the

---

[19] Dryden v. Bd. of Pension Commrs., 6 Cal.2d 575, 59 P.2d 104 (1925).
[20] Talbot v. City of Pasadena, 28 Cal. App.2d 271, 82 P.2d 483, 100 P.2d 806 (1938).
[21] Lund v. Minneapolis Fire Dept. Relief Assn., 137 Minn. 395, 163 N.W. 742 (1917).
[22] Hunter v. Quick, 183 Okla. 19, 79 P.2d 590 (1938).

resignation from the civil service commission will not act as a reinstatement so as to entitle dependents of a deceased fire fighter to a pension.

Upon leaving the service, a fire fighter is not entitled to a refund of the contributions or salary deductions which have gone into the pension fund in the absence of a statutory provision authorizing such a refund.

Ability to Perform Light Duty

A practice too frequently indulged in by many large fire departments is that of filling the fire prevention bureau, dispatchers office, fire academy, and other special duty assignments requiring highly technical skills, with crippled fire fighters who have neither the desire nor the qualifications for performing these important tasks. For instance, the Illinois pension statute defines a fire fighter as a member who performs "any full time position in the fire department," and this was held to include duties in the fire prevention bureau of the City of Des Plaines.[23]

A similar result was reached in the case of a Rockford, Ill., fire fighter who had his back injured in an off-duty accident and applied for a nonservice connected disability pension. It was denied, for even though he was incapable of performing the regular duties of a fire fighter, he was not incapacitated for other duties of a lighter nature, such as being a fire inspector. The pension statute did not require that a fire fighter be put on disability pension unless his injuries were such that it became necessary to retire him from the service, and in this case the plaintiff was offered an assignment as fire inspector at the same rank and pay as he had in the fire station, and it was conceded that he was able to perform such duties.[24]

In a New Jersey case involving a fire fighter who was thrown from the tiller seat on the way to a fire, the court ruled that he could be assigned to light duty in lieu of being granted a pension.[25] But an opposite conclusion was reached in a New York arbitration finding, where a fire fighter was offered a "regular" position in the fire prevention bureau after receiving injuries in the line of duty. After he refused to report to duty in the new assignment, preferring to obtain a disability pension, he was fired. In sustaining the grievance brought by the discharged fire fighter, the arbitrator observed the trend in the courts toward liberal interpretation of disability pension laws, and ordered the city to pay his full wages, including back pay.[26] Though a light duty assignment apparently satisfied the State Civil Service Department, a previous state trial court held that offering a job with "full duties" in the fire prevention bureau was not the same as a full job in the fire department as a whole, for the latter would require the services of a normal, healthy fire fighter.[27]

[23] Peterson v. Board of Trustees of the Firemen's Pension Fund, 296 N.E.2d 271 (Ill. 1973).
[24] Morse, H. Newcomb, "Legal Insight: No Pension for Back Injury," *Fire Command!*, Vol. 38, No. 1, Jan. 1971, p. 27.
[25] Selig v. Firemen's Pension Fund, 119 N.J.L. 266, 196 A. 204 (1938).
[26] City of Jamestown and Local 1772, AFL-CIO, Jamestown Professional Fire Fighters Association (Hyman, 1974), FIRE DEPT. PERS. REP., Jan. 1974, p. 3.
[27] Skorko v. City of Binghamton, Broome County Sup. Ct. (N.Y. 1971).

While it might seem reasonable to assign a fire fighter who has used up his sick leave to a special duty assignment pending the medical determination of whether or not he has fully recovered, as from a heart attack for instance, there is no unanimity of opinion that this action is either just or legal. Some pension laws provide that a fire fighter may draw his full pay for a year while recovering from a line of duty injury, unless he is granted a disability pension prior to the elapse of that period; however, it is the author's opinion, that when a medical panel finds that the stricken fire fighter's condition has stabilized, and it is determined that he is permanently disabled and unable to perform the full duties of a regular fire fighter, he should be retired and not kept on indefinitely in a special assignment *unless,* (1) his condition is such that he can perform such duties with no difficulty, (2) he has the knowledge and skills to handle the job, (3) he requests the assignment in lieu of retirement, and (4) the officer in charge of the bureau to which he requests duty approves the request.

Other Skills or Other Income

Unless a pension statute specifically makes such outside income or extraneous skills a factor in fixing the amount of a pension, it is improper for a pension board to base the amount of pension upon these considerations.[27]

Compulsory Retirement Age

A Kentucky statute requiring that fire fighters and policemen be subject to compulsory retirements according to the rules of the city in which they are employed was held not to require the adoption by a city of a compulsory retirement age. The court of appeals pointed out that the Kentucky Employees Retirement System, which serves state employees, does not have a compulsory retirement provision.[28]*

Accepting Other Employment

After receiving a retirement pension, a man's right to the pension may be suspended by charter provision upon reentering public employment. But in the absence of such legislation it has been held that reemployment is no ground for termination of a retirement pension. It would, however, be ground for revoking a disability pension where it is properly established that the nature of the new employment is such that one who is able to perform its duties could also perform the normal functions of a fire fighter.

---

[28] Terrill v. Barber, 515 S.W.2d 239 (Ky. 1974); Fire Dept. Pers. Rep. No. 5, May 1975, p. 4.

*The author has been present at pension board hearings when the fire fighter applicant, having been found permanently disabled and medically eligible for retirement by a medical panel, was being interviewed for the purpose of determining the amount of disability pension to be awarded. It was especially distressing when the fire fighter was asked whether or not it was true that his wife owned an apartment house that produced substantial income, and if so, how much, in the effort of the board to fix the percentage of his salary that would be awarded. Presumably, the more outside income that was available to the applicant, the smaller the percentage of his salary would be awarded as a disability pension. Similarly, the board was heard to ask a fire fighter if it were not true that he had graduated from college with a degree in accounting, and that, despite his being in a wheel chair, he might be able to accept employment as a bookkeeper after his retirement.

Age Limitations

In a case involving the Wilkes-Barre Fire Department, the Supreme Court of Pennsylvania was called upon to consider the involuntary retirement of two fire fighters. The court commented as follows:

"The city contends that it may, without violating civil service regulations, for the purpose of promoting efficient service, retire employees after 20 years of service. Our view of the case does not call for a decision on that precise point. Boyle v. Philadelphia (338 Pa. 129) decides that a city, notwithstanding civil service regulations, may provide for compulsory retirement of firemen at the age of 65, and it may well be that the rationale of that case will support compulsory retirement after 20 years of service. But, in either case, compulsory retirement of civil service employees at 65 or after 20 years' service can be accomplished only be a general, nondiscriminatory ordinance which operates uniformly and equally upon *all* the members of the class created by the ordinance. Where a bona fide attempt is made by a municipality to improve its police and fire service, *and all employees of the same class are treated alike,* it would seem that there can be no doubt that the municipality has the right to adopt a plan of demotion and retirement based upon age limitations. To say that for economy a police force can be reduced, but for efficiency men too old for real service cannot be retired, does not make sense. *This is not the case of an individual, it is the case of all patrolmen, hosemen, and laddermen of Philadelphia.* Our decision rests upon the particular facts here presented, involving a *general nondiscriminatory age classification.* It does not open the door to removals for political or partisan reasons, made under the pretense of economy or efficiency."[29] (Emphasis added.)

The court went on to say that if, in the formulation of its public policy, a city determines that after 20 years of service its employees have outlived their usefulness as active fire fighters, it follows that all fire fighters within that class are equally obsolescent. It cannot select two members of that class, retire them, and retain in active service 35 other employees in the same class. If one of a class is to be retired because he is a member of an ordained classification, all employees in that class must be retired. So long as others in the same category are retained in active service, appellants may not be retired. They can be discharged by, and in conformity to, the civil service provisions of the act. The court held that the resolution was invalid and said that the salaries of appellants could be recovered as damages in a mandamus action.

Disability Pensions

The determination of the question of whether a disability has ceased has been held in a California case to be a judicial function, and a failure to grant a hearing is ground for attack upon the determining board's finding.[30] How-

---

[29] Boyle v. Philadelphia, 338 Pa. 129, 12 A.2d 43 (1940).
[30] Sheehan v. Bd. of Police Comm'rs., 197 Cal. 70, 239 P. 844 (1925).

ever, failure to make an objection (that a hearing had been denied) for a number of years, where the decision of the examining board had not been erroneous, does not give rise to a right to compel the city to reinstate the pension.[31] The court said:

"It may be conceded for purpose of appeal that petitioner could not be deprived of a right to receive a pension by an erroneous determination, without a hearing. The failure to grant a hearing would then be ground for attack upon the proceedings of the board, as a result of which petitioner might be given a hearing to determine anew, and properly, the question of disability. . . . But surely that is the extent of his right. He neglected to avail himself of it for over three years, and refused to make his services available to the city when, by the terms of the Charter, he was no longer entitled to a pension. To permit him now to recover back payments, not because he is entitled to them, but because he was denied a technical right to a hearing, would be to grant a remedy against the city highly penal in its effect, and wholly without relation to the actual right claimed to be violated."[32]

A disability pension cannot be denied on unreasonable or arbitrary grounds. For example, where a San Antonio, Texas, fire fighter became unable to perform his duties because of an injury which caused his little finger to become stiff, he underwent an operation to correct it but it was unsuccessful. A medical panel then found that unless he were to have his finger amputated he would be totally and permanently disabled to carry on his duties; he refused to have the operation, and as a result, the pension board denied him a disability pension. The Court of Civil Appeals overruled the board's decision, as did the District Court, saying that the pension board had no authority to order such an operation and could not penalize him for refusing to have it.[33]

A pension board's denial of a disability pension which goes against the overwhelming evidence presented in support of it will also be overturned by the courts. For example, take the case of a Houston, Tex., fire fighter. His entrance medical examination showed that he was in good physical condition, and another examination upon his return to the fire department after two years of military leave also proved him to be in good physical shape. After being thrown thirty feet from the fire truck in a collision with an ambulance, his private doctor found that his back injuries had resulted in permanent disability. However, the city physician said that he was able to return to duty, and while loading wet hose onto a fire truck he again injured his back, causing him to fall and become unable to get up. Subsequent medical examination disclosed that his back was severely injured with accompanying atrophy of leg muscles, and even with a back brace he had severe pain. Because the city physician again examined him and said he was fit for duty, the pension board denied his application for a disability pension. In affirming the district court's ruling that the fire fighter was entitled to a pension, the court found that the evidence did not

---

[31] Hermanson v. Bd. of Pension Commrs., 219 Cal. 622, at p. 623, 624, 28 P.2d 21 (1933).
[32] See note 31, at pp. 623 and 624.
[33] Morse, H. Newcomb, "Legal Insight: Permanent Off-duty Injury," *Fire Command!*, Vol. 38, No. 2, Feb. 1971, p. 21.

reasonably support the retirement board's decision when viewing the record as a whole.[34]

Other Matters Affecting Pension Rights

Where pension funds are deposited in a bank which subsequently goes into receivership, amounts which are found in an open checking account are not accorded preference in behalf of the fire fighters. However, amounts on deposit in a savings account are considered held in trust by the bank and fire fighters are entitled to a preference over general creditors.[35]

Where the widow of a Pasadena, Calif., fire fighter brought a civil action against the city to obtain pension payments in accordance with the city charter provisions,[36] she did not file the suit until several years had elapsed from the time of her husband's death from pneumonia. The court held that even though the statute of limitations had run against part of the accrued payments, the city's general demurrer could not be sustained, for her action was not barred on the ground of laches, as laches would not apply to recent accruals. Her complaint was demurrable, however, as to the retirement board of the city, for her sole remedy against it was a mandamus action. The court also held that the city could not demur on the ground that prior adjudication had determined that her husband, who had got drenched at a fire within 100 days of his death, did not die as result of a line of duty injury, for such facts did not appear on the face of the complaint.

Other cases[37,38] have similarly determined that mandamus is the proper remedy to compel ministerial action, including the issuance of vouchers to pay a fire fighter's pension, the placing of one's name on the pension roll, or to compel the city to levy taxes to meet pension fund requirements.

Under a Wisconsin statute giving any member of any city fire department, serving twenty-two years or more, a pension upon retiring from the service, the court said one who has served the required time, though part of it as a volunteer and part on a salary basis, is entitled to a pension upon his discharge.[39]

Legislation requiring insurers to pay a percentage of gross premiums into a fire fighter's benefit fund has been upheld, but the city treasurer, as a mere custodian of the fund, is not the proper party to sue to recover such a tax.

Although pension boards are allowed considerable latitude in determining who shall be pensioned, they cannot retire and pension fire fighters to make room for political favorites. In Horlick v. Swoboda,[40] the Pension Board of Racine, Wisconsin had retired the chief, assistant chief and two captains involuntarily on the alleged ground that they were too old to be physically fit for duty.

---

[34] Morse, H. Newcomb, "Legal Insight: Total Disability for a Strong Man," *Fire Command!*, Vol. 38, No. 7, July 1971, pp. 28–30.
[35] Fehrman v. Sioux City, 223 Iowa 308, 271 N.W. 500 (1937).
[36] Talbot v. City of Pasadena, 28 Cal. App.2d 271, 82 P.2d 806 (1938).
[37] Bd. of Trustees of Firemen's and Policemen's Pens. Fund of City of Marietta v. Brooks, 179 Okla. 600, 67 P.2d 4 (1937).
[38] Lage v. Marshalltown, 212 Iowa 53, 235 N.W. 761 (1931).
[39] State v. Knowles, 145 Wis. 523, 130 N.W. 451 (1911).
[40] 225 Wis. 162, 273 N.W. 534 (1937).

In determining the amount of a retirement pension, the board or commission is bound by the local laws, and with respect to disability pensions, the amount of a disabled fire fighter's pension is determined by the loss of body functions which affected his work as an active fire fighter and not by percentage of industrial or economic impairment. In Foster v. Pension Board of Alameda,[41] a lieutenant in the fire department had been retired on account of an injury to his knee received in line of duty and the applicable pension provision was: "The amount of such compensation (under the Workmen's Compensation Act) and the amount allowed by the Board shall not in any event exceed one-half of the member's salary at the time of injury."

Widows and Dependents

Where a widow is entitled to a pension until she remarries, an annulment of a subsequent marriage will not entitle her to a restoration of her pension rights.

A failure of the agent of the pension board to make out a certificate of health after the employee's examination could not be a basis for denying a widow the benefits of the pension fund.

Under the charters of some cities a widow is not entitled to a pension after the death of a member who has been retired on a disability pension. But where a fire fighter was killed on duty after serving fifteen of the twenty-two required years for a service pension, the court favoring a liberal construction of the pension law, granted the widow a pension.

The word "children" in pension acts has been construed to include grandchildren of the deceased pensioner, (whether their parents died before or after his decease), and adopted children, in states where adopted children are given all the rights of natural children.

Where the court construed a Los Angeles City ordinance to preclude children of a deceased policeman over sixteen years of age from receiving a pension, the city was permitted to recover the amount from the children which had been improperly paid out of the pension fund.

Under a city charter requiring payment of pensions to a "child" or "children" of a deceased, retired member of the fire department until the child or children should attain age of eighteen years, the court held that quoted words referred to any child or children living on the member's death, and included a child who was born after the member's retirement. The court held also that a child who was born to the member and third party to whom the member was not married and who was legally adopted by the member as authorized by statute also was entitled to pension.

Likewise, the supreme court in the State of Washington held that even though the dependent children of a deceased fire fighter were in the legal custody of their mother who had divorced their father prior to his demise, they were entitled to the pension which the City of Everett provides for minor children.[42]

---

[41] 23 Cal. App.2d 550, 73 P.2d 631 (1937).
[42] Morse, H. Newcomb, "Legal Insight: Benefits for Surviving Children," *Fire Command!* Vol. 42, No. 2, Feb. 1975, p. 32.

Veterans and Employees on Military Leave

The courts have ruled that under certain charter provisions the time spent in military service during World War II counts toward a service pension, and that retirement may be taken while still on active military duty during or after release from active duty.

"Vested Right" as Applied to Pensions

To say that a pension is, or is not, a vested right is meaningless unless given a specific application to a given set of facts. A typical situation involving this question arises where a fire fighter is retired on a pension equal to one-half the salary of his position. If the city, in financial difficulties, subsequently reduces the salaries of fire fighters and then attempts to decrease the benefits to pensioners to a corresponding amount, the issue is immediately raised as to whether the retired fire fighter has a vested right to a fixed amount. The answer sometimes given to this question is found in a decision in a Texas case. In holding that a retired policeman does not have a vested right to unaccrued payments, but a mere expectancy which is not infringed upon by subsequent legislation allowing a reduction of payments, the court said:

"In our opinion, the contract entered into by the employee with the city is made subject to the reserved power of the Legislature to amend, modify, or repeal the law upon which the pension system was erected and this necessarily constitutes a qualification upon the anticipated pension and a reserved right to terminate or diminish it."[43]

The above case was cited with approval in a case in which the Puerto Rican retirement statute of 1923, providing for optional payments to a pension fund, was held not to give government employees a vested right to recover a pension but a mere possible contingent right in the future which disappeared when the statute was repealed in 1925 by an act creating a compulsory pension fund.[44]

Thus the general rule is that pension laws may be amended or repealed without impairing any vested right or obligation of contract, and that "as against the state or its political subdivisions, there is no vested right in a pension accruing from month to month." This is so, notwithstanding the fact that compulsory deductions may have been made from the employee's salary toward a pension fund, for the customary reason that "a pension is a bounty springing from the appreciation and graciousness of the sovereign and may be given or withheld at its pleasure."

Therefore, in determining questions of this nature, there are at least two issues which must be decided at the outset: (1) What is the right which is claimed to be vested, and (2) when does it vest, if ever?

Where the right existed to the beneficiary of a deceased policeman for a lump sum but the law creating such right was repealed nine days before his death, the Supreme Court of the United States, after holding that no obligation of contract had been impaired, set forth what has been called the "California

---

[43] City of Dallas v. Trammell, 129 Tex. 150, 101 S.W.2d 1009, 112 A.L.R. 997 (1937).
[44] MacLeod v. Fernandez, 101 F.2d 20 (U.S. C.C.A. 1st, 1938).

Rule": "Until the particular event should happen upon which the money, or part of it was to be paid, there was no vested right in the officer to such payment. His interest in the fund was until then a mere expectancy, created by law, and liable to be revoked or destroyed by the same authority."

And where the right is not to a lump sum, but to a monthly payment, in an amount equal to a fixed percent of the salary, future payments may be lawfully reduced, even after the employee has become eligible to retire on a pension, because the right which was vested was the right to have the pension, but not a particular number of dollars. And as a corollary to this proposition, it will be noted that under such circumstances, the pensioner is entitled to an increase in the monthly payments where the salary of his former position has been increased subsequent to his retirement. Nor does such an increase constitute an unconstitutional gift of public funds.

A California case ruled that where the voters had repealed the charter provisions relating to pensions that the city was obligated to provide some kind of a pension to a fire fighter upon retirement who had met all the conditions for the abolished pension system.

Judicial Determination of Pension Rights

It is sometimes said that a board or commission which has the power to determine whether new pensions shall be granted or old ones revoked, acts in a ministerial capacity. Such a determination has also been held to be a quasi-judicial function.

In the absence of a statute allowing recourse to the courts, on the question of whether the action of a board in refusing to grant a pension may be thus reviewed, the general rule is that the court will not substitute its judgment for that of the body on a matter within its discretion, as long as it does not appear that the board's action was arbitrary or capricious.

In a California case,[45] the court cites with approval a statement from an earlier case: "Hence, whenever any board, tribunal or person is by law vested with authority to decide a question, such a decision, when made, is *res adjudicata,* and as conclusive of the issues involved in the decision as though the adjudication had been made by a court of general jurisdiction."[46]

This does not mean, however, that the courts will not review the findings of such boards, for even if the legislature makes the board's finding conclusive, yet such a proceeding must conform to the ordinary standards of fairness and reasonableness, and its decision is reviewable to show that the denial of the pension was in excess, rather than in the exercise of the board's authority. The California Supreme Court found that a board had abused its discretion when it found that a policeman's death from a heart attack was not the result of an injury received in line of duty.[47] In Carleton v. Board of Pension Fund

---

[45] McColgan v. Board of Police Commrs., 130 Col. App. 66, 19 P.2d 815 (1933).
[46] Mogan v. San Francisco Bd. of Police Comm'rs., 100 Cal. App. 270, 279 P. 1080 (1929).
[47] Buckley v. Roche, 214 Col. 241, 4 P.2d 929 (1931).

Commissioners[48] it was held that under the statutes of Washington the Superior Court (on a writ to review a board's denial of a pension) had no right to take evidence and hear the case anew, but its authority in such a case is merely to determine whether the board had jurisdiction, whether it proceeded in the mode required by law, and whether in making its decision the board violated any rule of law to the prejudice of the appellant. It was also held, however, that where the board in question had denied the pension because of the existence of certain facts, that it was in error to refuse the applicant the right to introduce evidence to disprove such facts.

The court may also review the evidence on appeal to ascertain whether the board's decision is supported by the facts, and where the denial of a pension is without support of any evidence, the court may consider itself authorized to render such a decision as should have been rendered by the board.

For example, the City of New York contended that the ultimate responsibility for determining the cause of a disability and whether it was service-connected rests with the board of trustees of the pension fund, and, notwithstanding the certification by the medical board that applicants were so incapacited by a line of duty injury as to be eligible for accidental disability retirement allowances, the board denied the pensions. The appelate court, in reversing the lower court ruling, declared that even though the power is vested in the board to make such decisions, it must be exercised reasonably. It may not unfairly and arbitrarily reject the conclusions of the medical board. Since pension board members are not doctors, they should not ordinarily substitute their own judgment for that of physicians who have been charged by law with the responsibility for conducting medical examinations and for investigating the cause of such disabilities. The court then declared that the general rule is that the proceedings of an administrative board should disclose the basis for the administrative denial in order that the court may determine whether the decision was reasonable or arbitrary.[49]

Review Activities

1. Research your state's laws concerning injuries that occur in line of duty. Discuss with your classmates whether or not these laws differ from those mentioned in the text. Next, write a statement expressing your opinion as to whether or not you think such laws should be standardized for all states. Then discuss your statement with your classmates. Be prepared to defend your reasoning by presenting examples of why you do or do not advocate such standardization.
2. Explain whether or not there is obligation of contract involved in receiving a pension. Describe the role of the "California Rule" in this matter.
3. What is your opinion concerning whether or not a fire fighter who has been disabled in the line of duty should continue to be employed by the fire department in a "light work" capacity?

---

[48] 115 Wash. 572, 197 P. 925 (1921).
[49] Morse, H. Newcomb, "Legal Insight: Judgment of Disability," *Fire Command!*, Vol. 38, No. 6, June 1971, pp. 29–30.

(a) In what instances should a fire fighter be kept on?
(b) In what instances shouldn't a fire fighter be kept on?
4. Can a fire fighter's dependents receive compensation in the event of the fire fighter's suicidal death? If so, under what circumstances? If not, why not?
5. Explain the importance of the time factor in filing a claim for a pension.
6. A fire fighter's right to a pension is determined by whether or not the fire fighter is a member of the fire department. Either by yourself or with your classmates, formulate a definition of exactly what constitutes a fire department member. Then compile listings of those persons you feel should and should not receive fire department pension benefits.

Chapter Ten
SALARY AND COMPENSATION

Matter of Law or Contract

Insofar as fire fighters are considered public officers, the rules governing the compensation of municipal officers in general are applicable to them, but no special rules can be laid down as to the amount or conditions of payment without reference to specific ordinance or charter provisions regulating the matter. In the absence of a law creating the position and authorizing the payment of salary, or an express contract, a fire fighter cannot recover compensation for services performed.

Overtime

Where a governmental fire fighter works longer hours than those required by law, without requesting additional pay or a decreased work period, no compensation can be recovered for the excess work.

Fire fighters working for a private industry are entitled to be paid for the entire period of their twenty-four-hour shift, notwithstanding that part of the time may be spent in athletic games and amusements. Within the meaning of the overtime provisions of the Fair Labor Standards Act, the Circuit Court of Appeals for the 5th Circuit held that such idle time was "working time." However, city fire fighters are specialists, not laborers or mechanics, and are entitled only to straight pay for any time they put in over the regular day.

The Supreme Court took this view in refusing 81 Phoenix, Ariz., fire fighters $118,000. This amount was claimed in overtime pay for hours on duty between May 1, 1944 and April 30, 1945, when manpower was scarce because of the war.

As pointed out in Chapter One, in discussing federal laws that affect the fire service, if the amendment to the Fair Labor Standards Act is ruled by the Supreme Court to be constitutional as applied to public fire departments, then provisions of the act governing working hours and over-time will have to be met.

Where fire fighters in the city of Portsmouth, Ohio, were paid on a monthly salary basis, the court of common pleas held that they were not entitled to extra compensation for additional time spent on the job in excess of any specified number of hours prior to 1971, the date when the City adopted the forty-eight-hour week for fire fighters.[1]

Salary

Minimum Wage

Though it was held an unconstitutional interference with local government for the State of Kentucky to fix a minimum wage for fire fighters,[2] contrary decisions were rendered in Illinois,[3] Texas,[4] and Arizona.[5]

In 1971 the supreme court of Montana held that fire fighters and policemen are professionals, and therefore not covered by the state's minimum wage law. The court looked to the decisions relating to the Fair Labor Standards Act, as it is similar to the Montana Minimum Wage Act. Both laws exclude from their application "any employee employed in a bona fide executive, administrative or professional capacity. . . ." After finding that federal decisions supported the constitutionality of such a law, it based its reasoning on the exclusion of fire fighters from its provisions on the fact that the legislature had always treated them as a professional and distinct class of employees.[6]

In an early Missouri case where the city council voted the fire chief a pay raise, the court held that the city auditor could not refuse to pay the new amount merely on the assertion that it is an unconstitutional change of remuneration during a term of office, as the chief was not an officer within the meaning of the constitutional prohibition.[7]

Salary as Incident to Office

Whether a fire fighter is an "employee" or a "public officer" is a controversial question, and with respect to his right to receive the salary for his position irrespective of whether or not he is able to perform the duties, the courts, even in the same state, may not be in accord. In holding that a San Diego, Cal., fire fighter could recover full salary while confined at home with a cold, a California appellate court declared that his right to compensation is not by virtue of contract, express or implied, but exists, when it exists at all, as a creature of the law and as an incident to the office. "The salary is incident to the title to the office and not to its occupancy or exercise."[8] But in a later California case, where a Los Angeles fire fighter who had been adjudged insane was denied the right to receive any salary as incident to an office, the court referred to the previous Jackson case by saying:

---

[1] Morse, H. Newcomb, "Legal Insight: Working Excessive Hours," *Fire Command!*, Vol. 40, No. 10, Oct. 1973, p. 22.
[2] Lexington v. Thompson, 113 Ky. 540, 68 S.W. 477, 57 L.R.A. 775 (1902).
[3] Gramlich v. City of Peoria, 374 Ill. 313, 29 N.E.2d 539 (1940).
[4] Dry v. Davidson, 115 S.W.2d 689 (Tex. Civ. App. 1938).
[5] Lohrs v. City of Phoenix, 83 P.2d 283 (1939).
[6] City of Billings v. Smith, 158 Mont. 197, 490 P.2d 221 (1971).
[7] State *ex rel* Johnson, 123 Mo. 43, 27 S.W. 399 (1894).
[8] Jackson v. Wilde, 52 Cal. App. 259, 198 P. 822 (1921).

"We deem it unnecessary to discuss the conflict of authority on the question of whether firemen are 'public officers' rather than employees, since such a characterization does not solve the fundamental problem of the proper interpretation of the ordinance and the state law. We cannot agree with counsel for the appellant that the question is settled in this state . . . regardless of what the situation may be in other states or under the charter of San Diego . . . we are forced to the conclusion that under the city charter of Los Angeles appellant was not a public officer. While the cases so holding (that a fireman is an officer) are the majority in number, in our opinion the minority cases have the greater weight of authority."[9]

Agreements to Accept a Reduction in Salary

Closely related to the preceding problem, and also involving the question of whether fire fighters are "public officers" or "employees," is the constitutionality of agreements between the city and its fire fighters wherein the latter agree to accept less salary than that stipulated by law. An Oregon court in 1899 held that a fire chief who had agreed with the fire commission to work for less than the statutory salary could not later recover the differences in amount, on the grounds that the contract was executed, and it amounted to a gift to the city each month.[10] This "executed contract" theory was also followed in a California case,[11] while the doctrine of estoppel was the basis for denying a recovery in New York[12] and Tennessee.[13] With respect to the executed contract theory, the view it represents seems inconsistent with the accepted principal that a liquidated debt cannot be satisfied by acceptance of a part in lieu of the whole. It is also difficult to understand how the city has been prejudiced by the conduct of its employees in accepting less than the legal salary to the point where the equitable doctrine of estoppel can be invoked. Even where it may be argued that in reliance upon a promise to accept less salary the city withheld its right to discharge some of the employees to effect the necessary economies, yet its position is not irrevocably changed, for it still retains all its powers of removal. In Wisconsin, the fire fighters were not estopped from demanding their full salary; the court said, "The better rule is that where the law, as in this case, prohibits a municipality from meddling, during an officer's term of office, with the compensation he shall receive for his services, the doctrine of estoppel cannot be invoked to thwart it." The court said that cases where the agreement to take less salary was made before accepting the office might be distinguished, but even under such circumstances the court declared that it would not apply the estoppel doctrine.

Where a Georgia statute permitted a person to waive any rights established by law in his favor, it was held not to contravene public policy for fire fighters to accept a reduced sum, even though the salary ordinance had not been

---
[9] Mason v. Los Angeles, 130 Cal. App. 224, 20 P.2d 84 (1933).
[10] De Boest v. Gambell, 35 Ore. 368, 58 P. 72 (1899). (See generally: "Back Pay—Defenses to Claims of City Employees," report No. 69 of the Nat. Institute of Municipal Law Officers.)
[11] Bloom v. Hazard, 104 Cal. 310, 37 P. 1037 (1894).
[12] Hobbs v. Yonkers, 102 N.Y. 13, 5 N.E. 778 (1886).
[13] Steele v. Chattanooga, 19 Tenn. App. 192, 84 S.W.2d 590 (1935).

changed. (The fire fighters' pay checks had a written acknowledgment on the back that said the checks were in full settlement of the salary owed and that the deductions were made in accordance with payee's written request.)[14]

Where the economies were achieved through passing another ordinance establishing reduced salaries, an argument by fire fighters that their contractual relationship to the city was thereby violated was rejected by a New Jersey court, notwithstanding the fact that the original ordinance fixing salaries had not been repealed. The court said, "public office is not a contract between the incumbent and the state or any of its governmental agencies. In the office there is no property right," and added that the mere fact that fire fighters contribute toward their pension does not create any contractual relation with the city which would prevent a reduction in their salaries.[15]

Where a San Diego police chief contended that the reduction in salary amounted to an unconstitutional amendment of the city charter, the California court held that the city council was merely exercising a power conferred upon it by the same charter. After remarking that the chief could have resigned if he was dissatisfied with his lot, the court declared that "although the term and salary may be named in the charter, yet there is no contract for a stipulated time or price that is binding on the public."[16]

In a majority of jurisdictions, however, agreements to accept less salary than the amount fixed by law are held to be void as against public policy and public officers are not estopped from demanding their full salary. This is true, whether the agreement was entered into before, at the time of acceptance, or after employment. It is also true that acceptance of a lesser amount than that due is not an accord and satisfaction because the original agreement is void; nor is it a waiver of the right to the legal salary. Mandamus may be used to compel the payment of any salary due.

Recovery of Back Pay upon Reinstatement

Where a fire fighter has been wrongfully excluded from his position, he is entitled, upon reinstatement, to all the salary that has accrued, and the city cannot set off his earnings made while working in another city's fire department during the period of wrongful exclusion.[17] A similar result was reached by the District Court of Appeal of Florida, when the city of Hialeah was not permitted to set off the $10,000 profits a fire fighter and his wife earned in their business during the period for which he was entitled to back pay.[18]

Sick Pay

A California court held that a member of the fire department who was off duty as a result of stepping on a nail in his own backyard was entitled to his salary during the period of disability, for he was to be considered "sick" within

---

[14] Barfield v. Atlanta, 53 Ga. App. 861, 187 S.E. 407 (1936).
[15] Adams v. Plainfield, 109 N.J.L. 289, 161 A. 647 (1932).
[16] Coyne v. Rennie, 97 Cal. 590, 32 P. 578 (1893).
[17] Padden v. New York City, 92 N.Y.S. 926, 45 Misc. Rep. 517 (1904).
[18] Morse, H. Newcomb, "Legal Insight: Back Pay," *Fire Command!*, Vol. 39, No. 8, Aug. 1972, pp. 72–73.

the meaning of the charter sick-leave provision.[19] In two other California cases, conflicting decisions were handed down by appellate courts, one holding that a fire fighter who was off duty with a cold could recover his salary for the period of absence,[20] while the other court declined to follow this view in holding that a fire fighter who had been adjudged insane could not recover salary for the period of disability.[21]

Where a Michigan arbitrator in 1974 found that a Pontiac fire fighter had been ordered back to duty while he was still ill, the fire fighter not only got his sick pay but was awarded one-tenth of his bi-weekly salary as a penalty, for the fire chief could not order him back to work without making some attempt to learn the condition of his health.[22]

Assignment of Salary

In jurisdictions where fire fighters are considered public officers they cannot assign their unearned salaries, for it contravenes public policy and impairs the efficiency of the service. And deductions from salary toward a pension fund cannot be recovered where the employment is terminated before the eligibility requirements for the pension are completed.[23]

Pay Parity with Policemen

In a case involving the policemen's union and the City of West Allis, although the Wisconsin State Employment Relation's Commission pointed out that its ruling in imposing parity did not imply that police-fire parity agreements were either traditional or justifiable, it did say that such agreements were not rare nor limited to police-fire settlements. In noting that police and fire fighters do have similar working conditions, a Milwaukee fact finder came to a similar conclusion as in the City of West Allis decision, and recommended parity with back pay awards in 1974.[24]

Salary Agreements

On October 3, 1975, the California Supreme Court decided Glendale City Employees' Association, Inc. v. City of Glendale. In this case a writ of mandate was granted compelling the city to provide salary increases according to a memorandum of understanding adopted by the employees and the city. The question in this case was whether a memorandum of understanding, prepared pursuant to the government code, was binding upon the city once it was adopted by the city, even though the government code provides that such jointly prepared written memorandum of understanding shall not be binding. The court held that the memorandum was binding on the city once accepted by a vote of the governmental body. The interesting point here is that the court appears to have decided that such memorandums are not binding (as per the

---

[19] Doody v. Davie, 77 Cal. App. 310, 246 P. 339 (1926).
[20] Jackson v. Wilde, 52 Cal. App. 259, 198 P. 822 (1921).
[21] Mason v. Los Angeles, 130 Cal. App. 224, 20 P.2d 84 (1933).
[22] Fire Dept. Pers. Rep. (sample issue), 1975, p. 13.
[23] Richman v. City of Geneva, 292 N.Y.S. 397, 161 Miss. Rep. 572 (1936).
[24] Fire Dept. Pers. Rep. (sample issue), 1975, p. 10.

government code) only so long as the governmental body has not accepted, by a vote, that which is contained in the agreement.[25]

Collective Bargaining

There is an increasing trend toward determining salaries of fire fighters through the collective bargaining process. For example, a 1975 Utah law now permits Utah fire fighters to organize and negotiate on wages, hours, and conditions of employment, and in the event of a deadlock, have the disputes resolved by binding arbitration. Fire chiefs, assistant chiefs and deputy chiefs are prohibited from belonging to an employee organization under the act.[26]

In Oklahoma, the State Supreme Court held that the collective bargaining and arbitration acts were constitutional, but said the city was not required to meet and confer with union representatives with respect to salaries. However, once a city agrees to collectively bargain and accept the opinions of the arbitrators such matters become binding upon the parties.[27]

A 1975 study of salary agreements by the Bureau of Labor Statistics, covering over 150 cities with populations of 100,000 or more, shows that these contracts range in length from one year to thirty months, and often include increased fringe benefits, such as medical and dental insurance, uniform allowances, educational incentive pay, increased longevity, overtime and premium pay. (The June 1975 issue of *Fire Department Personnel Reporter* contains an analysis of such increases for sixty-one fire departments across the country.)

Workmen's Compensation

In order to eliminate the employer's common law defenses of contributory negligence, assumption of risk, and the fellow-servant doctrine, in suits brought against him by employees, workmen's compensation laws have been passed in most states which abolish the delay and inequitable conditions of the common law remedy. But the wording in the acts of different states is not identical. Even where the laws of two states are similar, the courts of one may consider fire fighters to be employees, so as to bring them within the terms of the act, and in the other, the courts may consider them as "public officers" and exclude them from its benefits. In describing the California Workmen's Compensation Act, Mr. Witkin, in his Summary of California Law, says, "It is a compulsory act, establishing in all except certain designated employments an exclusive system of compensation for injuries to employees arising out of and in the course of their employment not caused by their intoxication or intentionally self-inflicted, and resulting in disability or death. The liability so established is incident to the status of employment. It is neither in tort nor in contract, but may possibly be described as quasi-contractual in nature."[28]

---

[25] Glendale City Employees Assn., Inc. v. City of Glendale, 124 *Cal. Rptr.* 513 (1975); discussed in *California Fireman*, Dec. 1975, p. 10.
[26] Utah Senate Bill 190, discussed in FIRE DEPT. PERS. REP., No. 8, Aug. 1975, p. 3.
[27] Midwest City v. Cravens, 532 P.2d 829 (1975); FIRE DEPT. PERS. REP., No. 5, May 1975, p. 3.
[28] See Note 5, So. Cal. L.R. 441.

In a Connecticut case,[29] a fire fighter regularly appointed under the charter provisions was held not to be an "employee" within the workmen's compensation act, defining an "employee" as any person who has entered into or works under a contract of service or apprenticeship with an employer. Whereas in Minnesota, both policemen and fire fighters are entitled to the benefits of the compensation statute which provides that workmen and employees shall be construed to mean, among other things, every person in the service of the city, under any appointment or contract of hire.[30]

In determining whether a volunteer is an employee of a village so as to come within the workmen's compensation law protection, the Supreme Court of Illinois used the following test: "The fact that the claimant was compensated for his services, was subject to the control of the fire chief with regard to the manner in which the work was done, was furnished tools, materials and equipment by the village, and was subject to discharge by the village board, clearly supports the finding of the employer-employee relationship."[31]

In Michigan the compensation statute excepts from its operation "any official of the state, or of any county, city township, incorporated village or School District." A policeman was held to be an officer, and not an employee.[32] These cases, based upon different provisions, emphasize the necessity for examining the act of each state, and the cases interpreting it, to determine whether a fire fighter does or does not come within the scope of the protection offered.

In some states, fire fighters and policemen come within the terms of the workmen's compensation act by virtue of express inclusion of all employees of the state or its governmental agencies. In others, they are included by reason of their being classified by the courts as "employees." Yet some courts merely look to the intention of the parties entering the insurance contract, and if it is determined from the amount of the premiums being paid that fire fighters were meant to be included, they are covered, without deciding whether or not they are "employees" or "officers."[33]

In a 1917 case of an injury that occurred in Saginaw, Michigan, where a fire captain suffered a rupture while helping others to manipulate a sleigh in a fire house, the court held that while ordinary fire fighters could be classed as employees, a captain, as an officer in the fire department, was a city official within the employer's liability law.[34]

The fact that a fire fighter may be eligible to receive a disability pension, or benefits from his relief association, does not bar recovery of compensation under these acts, nor does it reduce the amount thereof.

However, a 1973 Idaho Supreme Court decision permitted the City of Boise to deduct workmen's compensation payments from fire fighters' retirement benefits on the ground that duplication of benefits should not be allowed

---

[29] McDonald v. City of New Haven, 94 Conn. 403, 109 A. 176 (1920).
[30] State *ex rel* Duluth v. Dist. Ct. St. Louis County, 134 Minn. 28, 158 N.W. 681 (1915).
[31] Village of Creve Coeur v. Indust. Comm., 32 Ill.2d 430, 206 N.E.2d 706 (1965).
[32] Blynn v. City of Pontiac, 185 Mich. 35, 151 N.W. 681 (1915).
[33] Frankfort Gen. Insur. Co. v. Conduitt, 127 N.E. 212, 74 Ind. App. 584 (1920).
[34] McNally v. Saginaw, 197 Mich. 106, 163 N.W. 1015 (1917).

in a system of wage-loss protection.[35] The California,[36] Louisiana,[37] New York,[38] New Jersey[39] and Michigan[40] courts have come to a similar conclusion. The U.S. Supreme Court has held social security benefits can be affected by Workmen's Compensation awards.[41]

There are some state statutes which specifically provide that all members of an organized fire department, whether paid or volunteer, are included within the protection of the workmen's compensation act, and some states provide this protection for a fire fighter even if he is injured fighting a fire while off duty, or on vacation anywhere in the state. There are state statutes which also bring anyone who is called upon to assist at a fire or other emergency within the coverage of workmen's compensation in case of injury, and there is authority for holding that a civilian who has been killed while assisting a peace officer is temporarily in the service of the city as a policeman, and therefore an "employee" within the meaning of the workmen's compensation act.

Compensable Injuries

Whether or not a fire fighter's injuries or ailments are compensable under the provisions of workmen's compensation acts in states holding that they are included within its terms depends upon the rulings in the individual states. On the question of whether an ailment occurred in the course of the employment, it is unsettled whether pneumonia or heart attack is a "line of duty" ailment.

Although there is a statute creating a presumption in California that common ailments (including hernia) are the result of conditions occurring in the line of duty, it is a rebuttable presumption. Retirement boards have not felt any obligation to accept the conclusions of workmen's compensation determinations on the question of whether an injury or illness is service-connected.

For the first time in over thirty-five years of interpreting the California Labor Code Section 3212, relating to the presumption that heart attack is a line-of-duty injury for fire fighters, the California Court of Appeals held that a city was liable for payment of temporary and permanent disability to a fire fighter who had heart trouble fifteen months after the date of his retirement.[42] The court said that the use of the disjunctive phrase in Labor Code Section 3212, applying the presumption of heart trouble which "develops or manifests itself" during employment, evidences a legislative intent to assign different and distinct meanings to the words "develop" and "manifest." Accordingly, annulment of a "take nothing" award rendered by the Workmen's Compensation Appeals Board on the application of a retired fire fighter for disability arising out of heart trouble was required, where it appeared that the board had con-

---

[35] C.F.B. v. City of Boise and State Insur. Fund, 95 Ida. 630, 516 P.2d 189 (1973).
[36] Lyons v. City of Los Angeles, 119 *Cal. Rptr.* 159 (Cal. App. 1975); FIRE DEPT. PERS. REP., No. 4, April 1975, p. 5.
[37] Paterson v. Baton Rouge, 309 So.2d 306 (La. 1975).
[38] Geremski v. Dept. of Fire of City of Syracuse, 357 N.Y.S.2d 975 (N.Y. 1974).
[39] Conklin v. City of East Orange, 135 N.J. Super. 1313, 342 A.2d 872 (1975).
[40] Johnson v. City of Muskegon, 232 N.W.2d 325 (Mich. App. 1975).
[41] Richardson v. Belcher, 404 U.S. 78 (1971).
[42] Soby v. Workmen's Compensation Appeals Board, 23 Cal. App.3rd 555, 102 *Cal. Rptr.* 727 (1972).

sidered only whether the trouble arose out of his employment and not whether it had developed during his employment. Since heart trouble might develop even though it did not manifest itself during employment, the court remanded the case to the board to get further evidence.

In a Michigan case, where a Muskeagan fire fighter died of pneumonia as a result of a drenching and exposure at a fire, the court held that he had not suffered an "accident" within the meaning of the compensation law.[43] Cases in Oklahoma[44] and Nebraska,[45] however, held that fire fighters died of injuries received in line of duty where their deaths were caused by perspiration and exposure while making inspections, resulting in pneumonia.

The Supreme Court of New Jersey reversed a lower court decision which denied a fire fighter compensation for injuries suffered while he was sliding into home plate in a softball game sponsored by a volunteer fire company. The employee was not required or compelled to play softball, but considered it one of his duties. The team was equipped and uniformed by the company and the scores were posted in the local newspaper. The court ruled that the act which compensates fire fighters applies, and that a softball game, with the uniforms and publicity, was an "exhibition" which brought it under the terms of the statute. The court said that the statute allowed recovery in cases where the employer as well as the employee receives benefits of the activity. The legislature intended liberal interpretation of recovery, because of the humanitarian purpose of the Act. A dissent called the volunteer company a social club.[46]

Though not a workmen's compensation case, the Supreme Court of South Carolina found a fire fighter not to be covered by an insurance contract entered into by his fire department. He was injured by a stray bullet while watching television in the fire station, and though the insurance protected volunteers "while on duty," it only protected paid men "while on duty at a fire."[47]

In Bricker v. Los Angeles, where a motorcycle officer was injured while chasing a violator off duty, a California court ruled that a fire fighter or policeman is on duty twenty-four hours a day, and if he is injured while performing his line of work, whether he is actually on duty or not, the city is liable for the costs of medical care and other provisions under the Workmen's Compensation law.[48]

The City of Syracuse, N.Y., was allowed to receive credit for the more than $12,000 in hospital and medical bills which had been paid out in a fire fighter's behalf by a Blue Cross-Blue Shield insurance policy, when satisfying a court judgment that he be reimbursed "for medical and hospital costs attributable to his heart condition, paid by him . . ." The court rejected the fire fighter's argu-

---

[43] Landers v. City of Muskegan, 196 Mich. 750, 163 N.W. 43 (1917).
[44] *In re* Benson, 178 Okla. 299, 62 P.2d 962 (1936).
[45] Elliot v. City of Omaha, 109 Neb. 478, 191 N.W. 653 (1922); the California Supreme Court *et,* in 1968, held that lung cancer was compensable.
[46] Cuna v. Board of Fire Commissioners (1964), Supreme Court of New Jersey.
[47] Morse, H. Newcomb, "Legal Insight: Watching Television Doesn't Count," *Fire Command!*, Vol. 41, No. 7, July 1974, p. 32.
[48] Bricker v. Los Angeles.

ment that even though the city had paid the premiums on the insurance, it was a fringe benefit guaranteed in a collective bargaining agreement, and that to allow the city credit for the payments would deprive him of equivalent wages due to him. The court said that if the city were not allowed such credit, it would amount to double payment to the fire fighter, and such a result is not desirable.[49]

Average Man Doctrine

A principle widely used in the State of New York in determining whether or not there is substantial evidence that an injury is compensable under that state law is referred to as the "average man doctrine." This doctrine is used in satisfying the substantial evidence rule of causality as an essential criteria for award purposes. Whether a particular event was an "industrial accident" is to be determined, not by any legal definition, but by the common-sense viewpoint of the average man.[50] For example, if the average man would believe, after hearing substantial evidence, that a person's heart trouble followed the ordinary course of wear and tear of life, then compensation would be denied.

A medical opinion, in order to be acceptable, must be independent of the ordinary wear and tear of life causation in establishing a causal relationship between an alleged injury and a disability. Yet a doctor's testimony may be disregarded if the evidence was contrary to belief of the average common-sense man in favoring a causal relationship between disability and the accident.

The average man believes that injuries can result from exposure to high temperatures, extreme cold, poison gases, smoke, great fatigue, etc. In a New York case, heat was accepted by the court as a contributary cause, thereby establishing the substantial evidence necessary in proving causal relationship.[51] The same court has held that a predisposition to occupational disease would be no bar to a compensable claim if it could be shown that the occupational conditions aggravated the underlying condition.

Proof of Claim

Although an award cannot be sustained where it is supported by hearsay evidence alone, yet other evidence need not establish the accident independently of the hearsay. In order to prove a claimant's personal injury in a manner to justify an award of compensation, it would be helpful if the following items could be established by reliable evidence:

(1) good health prior to the accident.
(2) physical effects of the accident on the claimant, as related by a witness present at the time.
(3) description of claimant's normal duties, so as to show that he was so engaged when injured.
(4) description of his activities on the day of the accident.
(5) injury report made by his employer.

---

[49] Geremski v. Dept. of Fire of City of Syracuse, 357 N.Y.S.2d 975 (1974).
[50] Masse v. J. H. Robinson Co., Inc., 301 N.Y. 34, 92 N.E.2d 56, at p. 57.
[51] Cottrell v. Pleasantville Fire District, 279 App. Div. 1124, N.Y.S.2d 588 (1952).

SALARY AND COMPENSATION

Review Activities

1. Discuss with your classmates your area's fire department policy on sick pay, overtime pay, recovery of back pay, collective bargaining, and salary agreements. Then write a brief statement explaining how they are similar to or different than those of the cities referred to in the text.
2. Describe the "average man doctrine."
3. Explain how a fire fighter's overtime pay differs from a laborer's overtime benefits.
4. In a general class discussion, elicit examples of compensable injuries in your state. Then describe in writing how your state's Workman's Compensation law on compensable injuries differs from those mentioned in the text.
5. In outline form, list the items a fire fighter should include in a claim for compensation.
6. Due to lowered economic conditions, some cities have found it necessary to lay off members of their police and fire departments. Write a brief description of your feelings concerning this matter, including in your description your thoughts on whether a fire fighter is an employee or a public officer.

Chapter Eleven
TERMINATION OF EMPLOYMENT

Reductions in Force

In the absence of a state statute to the contrary, a city can create or abolish positions within its government as it finds necessary or desirable. However, when the City of Utica, N.Y., laid off eight fire fighters who had been recently appointed to fill positions created by the city council, the Supreme Court of New York ruled that the power to create or abolish fire fighter positions rests with the city council by ordinance. Further, the refusal by a board of estimates or other branch of city government to fund such positions after being duly authorized is unlawful. Hence, the court ordered the authorized vacancies to be filled and the fire fighters who had been laid off were reinstated.[1]

To effect economies in municipal government is a valid ground for reducing the number of men in the fire force, and this may be accomplished by abolishing positions, or merely removing men from the payroll until further notice. And this may be done even where the position is under civil service in the absence of some legislative inhibition. It has been held unncessary, however, to abolish the office or position to save the city money, where the same result could be achieved by reducing all the battalion chiefs to the rank of captains.[2]

Removal or Suspension

There are no universal rules as to what constitutes proper procedure or sufficient grounds for the removal or suspension of fire fighters, but reference must be made to the controlling local laws, charter provisions, civil service regulations, and the rules of the fire department concerned. The powers of removal and suspension can be exercised only by the officers, board, or tribunal designated by law and in the manner pointed out by such laws and regulations.

In the case of Nelson v. Baker[3] the Oregon court said: "The powers of

---
[1] Timpano v. Hanna, 355 N.Y.S.2d 226 (1974); FIRE DEPT. PERS. REP. (sample issue), p. 8, 1975.
[2] Durkin v. Board of Fire Comm'rs., 90 N.J.L. 670, 101 A. 1053 (1917), aff'g 89 N.J.L. 468, 99 A. 432 (Sup. Ct. 1916).
[3] Nelson v. Baker, 112 Or. 79, 277 P. 301, at p. 303 (1924).

removal of an officer by the executive authorities are usually defined by the statutes upon which they depend. The executive power of removal may either be an arbitrary or a conditional one. If the power is an arbitrary one, no formalities, such as the presentment of charges or the granting of a hearing to the person removed, are necessary to its lawful exercise. A conditional or limited power of removal, as for cause, may, however, be exercised only after charges have been made against, and a hearing accorded to the person removed, in order to effect a valid exercise of the power."

Removal of Entire Fire Department

It has been held that an entire fire department may be ousted through repealing of the ordinance creating the fire fighters' positions; however, the mere act of placing a fire department under civil service was held not to create vacancies in the positions held at the time the act was adopted. Neither can a board of police and fire commissioners sidestep the rules relating to the removal of fire fighters by attempting to discharge the whole force and then immediately reappoint all but a few that are intended to be dropped. A Pennsylvania statute authorized certain classes of cities to reduce the work force; however, the city of Scranton did not fall in any of these classes. Yet, another statute prohibited a city from removing a fire fighter without his consent except for disciplinary reasons. Thus the court ordered twelve fire fighters who had been forced to retire for economic reasons to be reinstated with back pay to the date of dismissal.[4]

Even though a city has entered into a collective bargaining agreement which specifically excluded economy or abolition of a position as grounds for dismissal, a New York court held that such agreements could not deprive an appointing official of the power to abolish a civil service position when acting in good faith. Therefore the termination of employees purely for economic reasons was lawful.[5]

Seniority as a Factor in Layoffs

As a general rule, when layoffs are necessary, the principle of seniority is followed with the last hired to be the first fired. This rule was unsuccessfully challenged by black Detroit fire fighters in 1975 when 45 percent of the 302 black members were laid off due to a severe revenue shortage. In upholding the Detroit Fire Fighters Association's contention that layoffs should proceed according to seniority, U.S. District Judge James Churchill declared that the unemployed black fire fighters "want the court to give them seniority which they have not earned and which they have not been deprived of for any past reason."[6]

A California court reached a similar decision in the case of a Huntington Park fire fighter who had lost his position, along with others, in an economy move, in 1974. The plaintiff, with ten years of service, had more seniority than

---
[4] Bauer v. Peters, 331 A.2d 245 (Pa. Commonwealth 1975); FIRE DEPT. PERS. REP., No. 3, March, 1975, p. 8.
[5] Schwab v. Bowen, 363 N.Y.S.2d 434 (Miss. 1975); FIRE DEPT. PERS. REP., No. 3, March 1975, p. 8.
[6] "Employee-Employer Relations Developments," *California Firemen*, Sept. 1975, p. 7.

most of the members retained in the Department, and he won full reinstatement with all lost pay.[7]

Terminal Leaves

A growing practice in fire departments is to pay its retiring members for accumulated vacation and sick leave time that is owed them on the terminal date. Sometimes this is paid in a lump sum and sometimes on a post-retirement monthly basis, depending upon the ordinance, charter provision, or statute which creates this right. If the right was in existence at the time a member entered the fire department, its subsequent abolition prior to his termination of employment will not deprive him of it.

For example, where a member of the Roseburg Rural Fire Protection District had accumulated forty-seven days of sick leave prior to his termination, the Supreme Court of Oregon, in affirming the judgment of the circuit court, declared that the fire district's sick leave provision was a part of the inducement to accept employment, and therefore can be considered as a part of the plaintiff's contract which could not be rescinded after he had earned it.[8]

Ability to Perform Light Duty as a Bar to Removal

As previously noted in the subject of pensions (Chap. 9), the courts are not unanimous in holding that fire fighters who are capable of performing light duty, e.g., in the fire prevention bureau, cannot be retired on a disability pension.

Grounds for Removal or Suspension—Constitutionality

The mere fact that a fire chief, fire commissioner, board of directors, civil service commission, or other authority is permitted under the applicable law to prescribe the conduct of its members, does not authorize them to impose rules which violate constitutional rights. The contention that such rights were violated has been sustained sometimes but often rejected by the courts in various states when considering rules which prohibited residence outside the city, moonlighting, growing beards, joining subversive associations, running for office, by-passing "channels," posing for advertisement photos, talking to the press, failure to lose weight,[9] refusal to attend off-duty training classes,[10] etc. Some court rulings on these constitutional questions will be considered here, since these issues generally arise when a fire fighter has been suspended or removed for reasons which he feels are unjust or untenable under the safeguards afforded him by state or federal constitutions.

Notwithstanding a Texas statute making it lawful to belong to labor unions, it was held not unconstitutional to remove fire fighters who had joined such organizations in violation of the department's rules and regulations.[11] A Pennsylvania court said: "Any association which, on any occasion, for any purpose,

---

[7] Hagan, J., "A History of CSFA Benefits," *California Firemen*, Oct. 1974, p. 10.
[8] Morse, H. Newcomb, "Legal Insight: Unused Sick Leave," *Fire Command!*, Vol. 40, No. 2, Feb. 1973, 29.
[9] FIRE DEPT. PERS. REP., No. 6, June 1975, p. 5; rule upheld.
[10] Morse, H. Newcomb, "Legal Insight: Insubordination over Night Class," *Fire Command!*, Vol. 40, No. 12, Dec. 1973, p. 28; rule sustained.
[11] McNatt v. Lowther, 223 S.W. 503 (Tex. Civ. App. 1920).

attempts to control the relations of members of either the police or fire departments toward the city they undertake to serve, is . . . subversive of the public service and detrimental to the public welfare."[12] There are other cases in accord, all of which are collected in "Power of Municipalities to Enter into Labor Union Contracts," Report No. 76, of the National Institute of Municipal Law Officers.

In a California case it was held that a refusal to testify against one's self could not be a valid ground for dismissal of a policeman.[13] But in Pennsylvania, a refusal to answer questions when summoned by the county grand inquest was held to be conduct unbecoming an officer.[14]

Removal is improper, however, where the sole ground is violation of a rule against filing a civil suit without first obtaining permission from one's superior officer, because it contravenes the principle that the courts of justice shall be open to everyone.[15]

Removal of fire fighters and policemen has been effected without constitutional objection, where the basis for removal was violation of the department's rule against engaging in political activity, against being absent without leave, failure to pass the physical test for ladder work, moonlighting, and intoxication, although merely "taking a drink" is not necessarily "misconduct," where, for example, it is doctor's orders, to test evidence, or in one's own home.

A police department rule against "conduct unbecoming a member and detrimental to the service" was held to be so vague as to be unconstitutional on its face, by a U.S. Court of Appeals in the Seventh Circuit: Since the U.S. Supreme Court declined to review the decision, it would seem that similar regulations in fire departments are also invalid. In this case an officer had been reprimanded for sending a letter to Milwaukee's labor negotiator complaining of overtime matters.[16]

It is not insubordination for a fire fighter to complain of grievances to a higher authority, for this has always been one of the attributes of citizenship under a free government; nor will this charge lie against a policeman who deliberates a few minutes as to his legal course of duty.

When an Oakland, Calif. fire fighter was suspended sixty days for violating a rule against impeding the progress, welfare, discipline, efficiency, and good name of the department because a small amount of marijuana was found in his home, the constitutional question was raised that the rule was too broad and its meaning not precise. The fire fighter denied using marijuana and the Superior Court of Alameda County agreed with his contention that there was no evidence showing that his alleged misconduct had resulted in any adverse effect upon the department. His suspension was revoked and he was awarded back pay with interest.[17]

---

[12] Hutchinson v. Magee, 278 Pa. 119, at p. 120, 122 A. 234 (1923).
[13] Garvin v. Chambers *et al*, 195 Cal. 212, 232 P. 696 (1924).
[14] Sonder v. Philadelphia, 305 Pa. 1, 156 A. 245 (1931).
[15] State *ex rel* Kennedy v. Rammers, 340 Mo. 126, 101 S.W.2d 70 (1936); State v. Barry, 123 Ohio St. 458, 175 N.E. 855 (1931).
[16] Bluce v. Brier, 501 F.2d 1185 (7th Circ.) cert. den., 95 S. Ct. 804 (1975).
[17] Cann v. Civil Service Bd. of City of Oakland (Cal. Superior Ct. 1975); FIRE DEPT. PERS. REP., No. 9, Sept. 1975, p. 7.

Another case involving marijuana and an equally vague rule was one wherein a New York City fire fighter, on leave leading to his retirement, was fired and denied a pension because of conduct tending to bring reproach and to reflect discredit on the department. Here again, the constitutionality of the rule was not raised, but the court held that the evidence was insufficient to show that the fire fighter knew that the bag, found in the back seat of the car in which he was riding at the time of his arrest, contained marijuana.[18]

While removal may be had for "good of the service," if the charges are not supported by sufficient evidence to constitute "misconduct in office" or "neglect of duty" they will be dismissed. And where the dismissal is made on one ground, such as "keeping aloof from the others," the written charges preferred against the fire fighter must not be fabricated on another charge. Nor must the charges be too trivial; a mere technical violation of a department rule, such as making an erasure on the records, forwarding a letter of complaint directly to the mayor instead of through "regular channels," or granting an interview to the press, will not justify dismissal where the individual's record has otherwise been untarnished. A removal for "cause" must be based on a ground which relates to and affects the administration of his office and the rights of the public.

In a Los Angeles case, the court upheld the removal of a fire fighter for "physical inability to properly perform the duties required of all uniformed personnel . . . such being caused by illness not incurred in line of duty as determined by competent medical authority after thorough examination."[19]

Where the ground of removal is "insanity," this "does not," said a New York court, "refer to any temporary aberration of mind or delirium . . . it has in contemplation a permanent condition of insanity or mental unsoundness such as will render the member of the force unable or unfit to perform . . . duty."[20]

Necessity for a Fair and Impartial Hearing

While by law in some jurisdictions it is not required, the general trend of recent legislation is to grant the right of a fair and impartial hearing upon sufficient charges before removing or suspending fire fighters and policemen from their positions. By the great weight of authority a notice and hearing is necessary for a removal "for cause," and by the term "hearing" is meant a proceeding "quasi-judicial" or "quasi-criminal" in nature, in which all the findings must be based upon legal evidence. However, as an Oregon court points out, "Where charges are required to be made and a hearing given in order to effect their (employees') removal, such a hearing is not governed by rules applicable to strictly judicial proceedings."[21]

It is not necessary in disciplinary cases to produce the same quantity of evidence as would be required in a criminal trial to obtain a conviction. For

---
[18] Arrastia v. Lowery, 356 N.Y.S.2d 306 (App. 1974); FIRE DEPT. PERS. REP., No. 9, Sept. 1975, p. 7.
[19] Hostetter v. Alderson, 38 Cal.2d 499, 241 P.2d 230 (1952).
[20] Reiblich v. Copsly, 130 N.Y.S. 597, 71 Misc. Rep. 50 (1911).
[21] Nelson v. Baker, 112 Or. 79, 227 P. 301, at p. 303 (1924).

example, if a member is charged with "conduct unbecoming," or "conduct tending to bring discredit to the department," it is not necessary to prove facts which would show a violation of the penal code or constitute a crime under any ordinance or statute, but only some measure of proof that goes beyond mere surmise and speculation that an unlawful or reprehensible act was committed in such a way as to adversely reflect on the fire department's good standing and reputation in the community. Thus, where a fire fighter and female companion engaged in sexual relations on a public beach, thinking that they were alone in a secluded cove, and a deputy sheriff stopped his car to investigate the reason for a crowd of spectators gathered on the brink of the cliff overlooking the beach, the fire department's trial board was probably justified in accepting the opinion of the sheriff, who arrested this oblivious couple, that such conduct was unbecoming a fire fighter, though he was off duty and out of uniform at the time of the incident.

Where a fire department captain was suspended for a month for being drunk while on duty, and after a hearing the penalty was reviewed and affirmed by the city's common council, an attempt was made to annul the council's findings on the grounds that the fire chief's uncorroborated testimony did not constitute sufficient evidence to support the charge. The city contended that the testimony of a doctor who had examined the captain several hours after he was first observed by the chief was sufficient corroboration. In affirming the city council's determination, the Appellate Division of New York found the evidence sufficient to conclude that the captain was under the influence of intoxicating liquor while on duty. The court said that all that was involved in the hearing was questions of fact, and it had no power to substitute its own views on a question of fact when that determination is supported by substantial evidence. There is no requirement, the court said, that the chief's testimony be supported by corroborating evidence, for determinations may be sustained solely upon the testimony of the complaining witness where such testimony is not shown to be incredible or impossible as a matter of law.[22]

With respect to a right to hearing, this is generally established in civil service rules, charter provisions, or grievance procedures, as a necessary ingredient of "due process" and a condition precedent to suspension or removal of any member who has completed his probationary period of service. However, the failure to demand a hearing within the time required by law may subsequently bar the right.

Up until 1975, in the case of a suspension of less than thirty days, some courts have held that a hearing prior to a suspension was not required to satisfy due process. But the new rule, as set forth by the U.S. Supreme Court in a case holding that students could not be suspended without a prior hearing,[23] was followed by the Seventh Circuit U.S. Court of Appeals in Muscare v. Quinn.[24] In this case a Chicago fire department lieutenant was suspended twenty-nine days for refusing to shave off his goatee, and the U.S. District Court said the

---

[22] Morse, H. Newcomb, "Legal Insight: Under the Influence... Judgment," *Fire Command!*, Vol. 39, No. 1, Jan. 1972, p. 30.
[23] 95 S. Ct. 729 (1975).
[24] 520 F.2d 1212 (7th Cir. 1975); Rhg. den. 1975.

lack of a prior hearing did not violate due process. In reversing this decision, the U.S. Court of Appeals declared that due process (under the 14th Amendment) requires that public employees be granted a hearing prior to suspension where they may be fully informed of the reasons of the proposed suspension and where they may challenge their sufficiency.[25]

Although the court does not spell out the nature of such presuspension hearing, it would appear that no formal adversary proceeding is contemplated, but merely an informal presentation of the charges being brought and a chance for the employee to explain or justify his conduct.

Right to Resist Dismissal by Improper Person

The applicable law usually specifies who has the authority to remove fire department personnel. It may be the fire chief, the fire commission, or other head of the department, or, after appeal, the civil service commission or municipal governing council.

In a case where the city adminstrator of the city of Fort Smith, Ark., dismissed the fire chief, the chief's protest to the civil service commission was to no avail. He then appealed his case to the courts, contending that the city administrator had no authority to fire him, but only the city's board of directors could do so. The Supreme Court of Arkansas held that since a state statute provided that fire fighters under civil service must be given written notice of the cause of their discharge, and since the applicable regulation of the civil service commission required such notice to be given by the head or governing body of the city, then only the city's board of directors could give such a notice, for in a city-administrator form of municipal government, the governing body is the board of directors. In this instance, neither the statute nor the regulation was complied with and the attempted discharge of the fire chief by the city administrator was unlawful.[26]

Right to Counsel at Hearing

In establishing a hearing procedure, such as a board of rights, city charters or ordinances often provide that the accused member may select any officer of the fire department to act as his defense counsel, or he may have a union representative or attorney assist him with the presentation of his defense.

A Seattle fire fighter was charged with incompetency and inefficiency for failure to operate a hydrant properly at a fire, and the fire chief followed the trial board's recommendation that he be fired. The fire fighter appealed to the civil service commission which then made an independent investigation and upheld the chief's action. In answering his objections, on appeal to the Washington Court of Appeals, the court said that whether or not he was denied counsel or the opportunity to cross examine his accusers, or call his own witnesses was irrelevant because the civil service commission had made an independent fact-finding effort which afforded him full due process.[27]

---

[25] FIRE DEPT. PERS. REP., No. 7, July 1975, p. 10.
[26] Morse, H. Newcomb, "Legal Insight: Who Fires the Chief?," *Fire Command!*, Vol. 40, No. 9, Sept. 1973, pp. 66, 67.
[27] Deering v. City of Seattle, 520 P.2d 638 (Wash. App. 1974); FIRE DEPT. PERS. REP. (sample issue), p. 10.

However, where an ambulance driver in the Los Angeles Fire Department was terminated following a hearing before the civil service commission, at which his request for a continuance pending the arrival of his attorney was denied, the California Court of Appeals held that the hearing was unfair; a civil service commission is bound by the same fundamental principles of fairness as are the courts, and a trial must not only be fair in fact but also in appearance. This was neither, and the court referred the matter back to the commission for a full hearing on the merits, with the presence of his attorney.[28]

A number of cases have held that an accused employee is entitled to have a union steward, official, or representative with him during an investigating interview conducted by the employer because of the rights afforded under the National Labor Relations Act. Though the NLRA is not applicable to public employers, yet fire departments having collective bargaining agreements may be affected by the latest Supreme Court decision affirming this right.[29] Thirty three states have adopted public employee relations acts or have promulgated agency or executive orders which are identical or similar to the federal statute.[30]

Right to Remain Silent

The U.S. Supreme Court has held that the constitutional right to refuse to be a witness against one's self in a criminal case is one of the most valuable prerogatives a citizen has, and this choice to speak out or remain silent is extended to public employees. Hence, a fire fighter may not be discharged for failure to give an account to his superiors for matters not directly related to the performance of his official duties. For example, in Cox v. City of Chattanooga,[31] a fire captain's employment was terminated because of his refusal to answer questions about a person suspected of murder whose name was found in his address book, and the court declared his discharge was illegal and in violation of his constitutional rights.[32]

Polygraph Test

Likewise, in the case of polygraph tests, where the subject under investigation does not relate to the fire fighter's duties, he cannot be forced to submit to such a test. In a Texas case,[33] a bare majority of the Texas Supreme Court ordered the reinstatement of a fire fighter who had been fired for refusing to comply with an order of his chief that he take a polygraph test following his arrest for receiving and concealing a stolen pickup truck. Though policemen can be discharged in Texas for refusing to take such a test, the court noted a distinction between law enforcement officers and fire fighters. However, other states may be more inclined to agree with the four dissenting judges who felt that the question of honesty in the fire service, where fire fighters have ready access to valuables in homes and stores, as well as in the fire station, should be

---

[28] Wood v. City Civil Service Comm'n, 45 Cal. App.3rd 105, 119 *Cal. Rptr.* 175 (2d Dist. 1975).
[29] National Labor Relations Board v. Weingarten, 95 S. Ct. 959 (1975).
[30] FIRE DEPT. PERS. REP., No. 6, June 1975, pp. 6–8.
[31] Cox v. City of Chattanooga, 516 S.W.2d 94 (Tenn. App. 1974).
[32] Morse, H. Newcomb, "Legal Insight: You Have the Right to Remain Silent," *Fire Command!*, Vol. 42, No. 10, Oct. 1975, pp. 36–37.
[33] Talent v. City of Abilene, 508 S.W.2d 592 (Tex. 1974); FIRE DEPT. PERS. REP. (sample issue), 1975, p. 11.

given the same consideration as honesty in policemen. The better rule would seem to be the same as that applied to police officers, i.e., "you have the constitutional right to remain silent, but you do not have the constitutional right to be a policeman!"

Right of Judicial Review

Judicial review of a dismissal action is denied in some jurisdictions on the ground that the action sought to be reviewed is ministerial in character, and not a proper subject for review. However, where the power of removal or suspension is conditioned upon the granting of a hearing to determine the truth or falsity of specific charges, the question of regularity of such proceedings is always open to judicial review.[34] This is true even though under specific provisions of the city charter the decision of the officer or tribunal is made conclusive.

Many courts limit their judicial inquiry to questions of whether there has been a clear abuse of discretion, sufficient evidence to furnish a substantial basis for removal, the proper carrying out of formalities, sufficiency of charge to authorize removal, or matters purely of law and jurisdiction. However, there are decisions which hold that the court has the right to see if there is sufficient evidence to support the charge and to review the merits of the case. In one case, the Illinois court found that the evidence was insufficient to enable the civil service commission to find the charges true, and declared that on certiorari the court had the right to examine both the findings and the evidence to determine: (1) whether the commission had jurisdiction, (2) whether it exceeded its jurisdiction, (3) whether there was any (or sufficient) evidence tending to support the charge, and (4) whether the proceedings were conducted according to law.[35]

Where a fire lieutenant was dismissed for conduct unbecoming an officer two years after committing the offense, the Appellate Court of Illinois held that such delay was not a bar to his discharge, where he had remained on the force, collecting his salary, and there was no loss of evidence which prevented him from presenting a defense that he had not, as charged, had sexual relations with a female at his fire station. Moreover, the court declared that the findings and conclusions of the commission on questions of fact are deemed prima facie true and correct.[36]

Probationary Employees

The dismissal of probationary employees, unlike that of permanent civil service employees, may be exercised under a power of summary removal, and the right to notice and hearing is not usually accorded until a permanent status has been acquired. Where the city charter does not set forth any specific

---

[34] See Note 21.
[35] Murphy v. Houston, 250 Ill. App. 385 (1928).
[36] Monroe v. Civil Service Commission of Waukeagon, 55 Ill. App.2d 354, 204 N.E.2d 486 (1965).

grounds for the discharge of a probationer and does not authorize the civil service commission to pass upon the sufficiency of the reasons assigned, it is clear that "the reason for the discharge of a person holding a probationary position is within the sound discretion of the appointing power."[37]

Demotion

As a general rule, a fire fighter who has received a promotion, and is found by his superiors to be doing unsatisfactory work, can be demoted without having charges preferred against him as long as he is still serving his probation in the higher position. A case which illustrates this principle is that of a fire captain who was demoted to sergeant during his probationary period. It was contended that a statute which permits a public employer to reduce a probationary employee to his former rank without giving written reasons is unconstitutional, arbitrary, and capricious. The appellate court ruled that the law was valid and said that the essence of probationary employment is that the employer has unfettered discretion in deciding whether to retain a probationary employee.[38]

Though courts in other jurisdictions (First & Seventh Circuits), in following the U.S. Supreme Court decision in Roth v. Board of Regents,[39] have reached a contrary conclusion in granting certain basic rights to probationary employees, the Kentucky Court of Appeals decision, discussed above, conforms to an earlier ruling of the Sixth Circuit Court of Appeals.[40]

Where the person demoted has already completed his probationary period, it is customary to follow whatever hearing procedure has been established by law or bargaining agreement in the presentation of charges and giving the person a chance to defend himself. Unlike the case of a probationary employee, an opportunity for appeal to higher authority is usually provided for the demoted member.

For example, where a fire captain was demoted to lieutenant following a hearing before the board of fire commissioners in which he was found guilty on three different charges, only one charge, that of making derogatory remarks about an acting captain, was sustained by the arbitrator. He was reinstated to his former rank and awarded back pay, but "fined" one week's pay as punishment for failing to have needed respect for his fellow officers.[41]

As noted in the preceding case, where the punishment imposed is considered to be excessive for the offense committed, the higher authority which hears the appeal may modify the penalty to bring about a fair result. Another illustration of this type of situation was the case of an assistant fire chief who was demoted to captain because of an accident caused by his running a stop light while responding to an emergency with all his warning devices in oper-

---

[37] Newald v. Brock, 12 Cal.2d 662, 86 P.2d 1047 (1939).
[38] Louisville Professional Fire Fighters Assn. v. City of Louisville, 508 S.W.2d 42 (Ky. 1974).
[39] Rath v. Board of Regents, 408 U.S. 564 (1972).
[40] Orr v. Trinter, 44 F.2d 128 (1971); FIRE DEPT. PERS. REP., No. 2, Feb. 1975, p. 13.
[41] FIRE DEPT. PERS. REP., No. 10, Oct. 1975, p. 5.

ation. Because he had previously violated the speed limits imposed by department regulations, he had been temporarily suspended by the fire chief and ordered to observe all traffic regulations even when driving to a fire. On appeal to the civil service commission, it found that the charges of neglect of duty and insubordination were correct, but reduced the penalty to a ninety-day demotion in rank. A further appeal to the circuit court resulted in an affirmation of the commission's decision, and this finding was then upheld by the Michigan Court of Appeals, which stated that it was not an abuse of discretion on the part of the commission to review the proofs of the alleged charges, as well as to review the reasonableness of the penalty, and to impose the lesser penalty of temporary demotion.[42]

Personnel Included in Removal Safeguards

While it has been held that the secretary to the fire commission does not come within the scope of the protection afforded by the city charter with respect to removal of members of the fire department, yet a fire department physician, veterinary surgeon, and upholsterer have been held to be entitled to the same rights accorded to regular fire fighters in the matter of removal.

Reinstatement

Where a fire fighter has been wrongfully suspended or removed from his position he may usually invoke mandamus to compel restoration. In some jurisdictions, however, he may not have the right to be restored to the same rank or position, and in others to only the difference in pay between the amount earned during the period of wrongful exclusion and the fire department salary. Mandamus to compel restoration of back salary and reinstatement should be brought promptly after the wrongful exclusion in order to avoid a charge of laches (undue delay) which might bar relief. (Although with respect to a statute of limitations on claims of money against the city it has been held that it does not start to run until after the mandamus proceeding.) In a Montana action for reinstatement and accrued salary started two years after ouster, it was held the six-months statute of limitations had not started to run and laches had not been pleaded by the city.[43]

The California Supreme Court, in 1964, compelled San Francisco's Civil Service Commission to reinstate a deputy sheriff who had been terminated because he filed a declaration of candidacy for election to the office of sheriff of San Francisco in violation of a city charter provision. The court held that this restriction of a fundamental right was unconstitutional. (The new Califor-

---

[42] Morse, H. Newcomb, "Legal Insight: Suspension of Assistant Chief's Driving Privileges," *Fire Command!*, Vol. 38, No. 5, May 1971, pp. 28, 29.

[43] Sweeny v. Butte, 64 Mont. 230, 208 P. 943 (1922).

nia government code provisions adopted in 1963 do not restrict public employees from running for office, whether local or partisan.)

In another case, decided the same day, the court held that an Alameda County charter provision barring political activity on the part of county employees was also unconstitutional, and required the reinstatement of a doctor who had been terminated for participating in a political campaign by becoming chairman of a speaker's bureau in a gubernatorial election.

A California case[44] held that a fire fighter who resigns under duress must file a claim for reinstatement within the same period as allowed in cases of unlawful suspension, lay-off, or discharge; a dictum added that there can be no reinstatement after a voluntary resignation.

Where a civil service employee had been dismissed for conduct unbecoming an officer because of his indictment on burglary charges, the fact that he was later acquitted was held by a Georgia Court of Appeals to be not controlling in the matter of his reinstatement.[45] The court agreed with the civil service board, which had conducted the hearing, that it is not necessary to prove that a man is a criminal to show that his conduct was unbecoming an officer. The reason is that criminal courts and administrative boards have different yardsticks for deciding upon a man's qualifications. A fundamental difference is the type of evidence required to "prove" a case.

In a criminal court, the prosecution must prove the defendant "guilty beyond a reasonable doubt," whereas board hearings are more like civil trials where "a preponderance of evidence" governs the outcome.

In this regard, boards and administrators generally are acutely aware of public relations problems. Any governmental service is hit hard and suffers a lowering of public regard if it fails to take a strict course in its standards of employee conduct. Therefore, no board could possibly accept the defense-oriented rule of "beyond reasonable doubt."

Review Activities

1. When does a fire department have the right to regulate a fire fighter's hair style? When does a fire department not have the right to regulate a fire fighter's appearance? Can not conforming to such regulations constitute grounds for removal or suspension? Explain.
2. Explain the rules or laws of your community's fire department concerning removal or suspension of its members.
3. Write brief explanations for each of the following:
    (a) What role does the economy play in the termination of fire department employees?
    (b) What role does seniority play in the termination of fire department employees?
4. If a fire fighter is appointed to sergeant and shortly thereafter is demoted to fire fighter, does that fire fighter have the right to appeal the decision?

---

[44] Moreno v. Cairns, 20 Cal.2d 531, 127 P.2d 914 (1942).
[45] Roberson v. City of Rome, 72 Ga. App. 55, 33 S.E.2d 33 (1945).

Why or why not? What factor determines a demoted member's ability to appeal?
5. Do you feel that a fire department member should be made to take a polygraph test? Why or why not?
6. Determine the constitutionality or unconstitutionality of each of the following hypothetical reasons for the termination of a fire department member:
    (a) Endorsement of a smoke detector product.
    (b) Drinking at the fire fighter's residence.
    (c) Long hair or a beard.
    (d) An arrest before fire department employment.
    (e) Inability to perform normal fire department duties.
    (f) A mental disorder, such as a nervous breakdown.
    (g) Sexual relations in public.

Chapter Twelve
DUTY OWED BY PUBLIC TO MEMBERS OF FIRE DEPARTMENT

Owner-Occupier's Relationship to Fire Fighters

In order to determine whether a fire fighter has a right to recover damages from the owner or occupier of property upon whose premises he has been injured, it must be first established that there has been a breach of duty owed to him. The nature of this duty depends upon his legal relationship to the possessor of the premises. In most jurisdictions he has been held to occupy the status of a licensee, since he enters the land of another under a right conferred by law and in the performance of a public duty. A licensee assumes the risks that may be encountered on the premises, and can only recover for wilful or wanton injury; but where he is on the premises with knowledge of the licensor, a duty to exercise due care in activities exists on the part of the latter, and a licensee may recover for active negligence or an "overt act of negligence." The *Restatement of Torts* applies the same rules to fire fighters, who are said to enter under privilege, as it does to other licensees, and imposes liability not only for failure to carry on activities with reasonable care, but for failure to warn of latent dangerous conditions on the land known to the possessor.[1]

The general rule is well set forth in a Tennessee case: "When firemen or policemen in the course of their duties go upon the premises of an individual, the latter owes no duty to them, except to refrain from inflicting upon them a wilful or wanton injury.... Firemen are authorized by law to go upon the premises of anyone in the discharge of their duties. The owner cannot prevent their entry, nor can he control their actions while they are there. They are accordingly mere licensees when acting within the city limits."[2]

The earliest case in this field, holding that a member of an Illinois fire insurance patrol was not intended to be protected by an ordinance requiring elevator shafts to be guarded, merely cites *Cooley on Torts* for its authority.[3] In a

---
[1] *Restatement of Torts*, Sections 341, 345, and 346.
[2] Buckeye Cotton Oil Co. v. Campagna, 146 Tenn. 389, 212 S.W. 646 (1922).
[3] *Cooley on Torts*, First Ed., 313 (1893).

Michigan Law review article the writer, in commenting on Judge Cooley's work, says:

"He does not distinguish between bare licensees and business invitees . . . it may be seen that he uses the term 'licensee' in a broad sense, covering the modern meaning of both 'business invitee' and 'bare licensee.' His paragraph on the right of a fireman, therefore, is not authority for holding that, as contrasted with those of an invitee, the rights of a fireman are those of a bare licensee. It is only authority for saying that a fireman enters as of a right and does not decide whether the right is that of what is now called a licensee or of an invitee. The many courts which have cited and followed Cooley as authority for a decision that a fireman is a bare licensee seem to have been under a misconception as to the sense in which he used the word."[4]

Are Fire Fighters Invitees?

The author of the above article offers a rather convincing argument that a fire fighter should be classified as an invitee: "If we accept the benefit to the land owner as the determining factor of an invitee, and it is generally so conceded, it is difficult to see on what basis, aside from *stare decisis,* it can be held that a fireman or a policeman is a mere licensee." Surely there can be no doubt that a fire fighter who enters a building to extinguish a fire, or to prevent a fire in an adjacent building from spreading, is performing a beneficial service to the owner and occupant.

Nor is this an isolated view. In a Missouri Law Review article, "Negligence —Liability of Possessor to Governmental Employees," the author recommends that the courts of Missouri give all governmental employees the status of business visitors, finding no valid ground for excluding firemen and policemen.[5] Professor Harper, in his work on *Torts,* has adopted the view that policemen and fire fighters should be accorded the status of invitees. Though admitting that the question is not settled, he adds:

". . . the weight of reason seems to be entirely with the latter view (i.e., that view which regards them as invitees). Such a person enters by reason of a social duty that the law should protect. All occupiers of land should share the burden incident to the discharge of these obligations by public employees."[6]

Although Professor Bohlen, in a Pennsylvania Law Review article, "The Duty of a Landowner Towards Those Entering the Premises of Their Own Right,"[7] is reluctant to take a stand which would impose too great a burden on the landowner, he does not take issue with the decision reached in the leading New York case of Miers v. Fred Koch Brewery which held the landowner liable to a fire fighter for injuries sustained when he fell into an unguarded hole in a driveway;[8] the court there held that as to that portion of the premises, at least, the fire fighter was owed the duty ordinarily accorded an invitee. The result, limited as it is to a prepared means of access over which those

---
[4] 35 Mich. Law Rev. 1157, "Torts—Are Firemen and Policemen Licensees or Invitees?"
[5] Mo. L. Rev. 23:69 Jan. 1958.
[6] *Harper on Torts,* Section 96, pp. 224, 225 (1933).
[7] 69 Univ. of Pa. Law Rev. 142, 337, 340, at p. 351 (1921).
[8] Miers v. Fred Koch Brewery, 229 N.Y. 10, 127 N.E. 491, 13 A.L.R. 628 (1920).

entitled to enter may be expected to use, is nevertheless a qualification of the statement found in *Pollock on Torts:* "He (the fireman) must take the property as he finds it, and is entitled only not to be led into danger, 'something like fraud.'"[9] Also, in view of the other classes of government employees who have been designated invitees, there appears to be no justifiable ground to deny fire fighters the protection ordinarily provided for invitees.

Are Fire Fighters Licensees?

A brief excerpt from the New Hampshire case of Smith v. Twin State Gas and Electric Co. sums up the current trend of thought in this field:

"It is difficult to perceive how the landowner is entitled by force of the law as to licensees, to protection from liability for defective conditions to a fireman, who enters his premises for the purpose of giving him protection. And there is much force in the reasoning in the Meirs v. Brewery case, which holds that firemen are neither licensees nor invitees of the owner of the premises on whose property they are fighting the fire, their right of entry being given by law regardless of the owner's attitude about it. And it is held immaterial whether firemen go to a fire in response to an alarm given by the owner of the property on fire or by someone else. The owner's act in sending the alarm does not make him an invitor in the legal sense, since it is not an invitation from him that is accepted, but a call to duty to which response is made (citing Lunt v. Post Printing etc. case).

"It is contemplated that he shall fight all fire however caused and run the risk of injury from whatever may be expected to be encountered in so doing. But in reason it is not to be contemplated that he shall have no protection against dangers not fairly to be expected or not incidental to his work."[10]

No distinction is drawn between the status of a fire fighter who enters to fight a fire on the premises entered, and one who enters in order to fight more advantageously a fire in adjoining premises.[11]

Perhaps the landmark case in Illinois is Dini v. Naiditch.[12] In overruling the earlier case of Gibson v. Leonard,[13] the appellate court held that fire fighters confer upon landowners economic and other benefits which are a recognized basis for imposing the common-law duty of reasonable care. Further, an action will be against the landowner for failure to exercise reasonable care in the maintenance of his property resulting in injury or death, due to the negligence of the owner or operator of the premises, to a fire fighter rightfully upon the premises fighting a fire. In this Illinois case, a fire captain was killed and a fire fighter was so seriously burned as to become permanently disabled when a stairway collapsed under them while fighting a fire in an old four-story brick building in Chicago. The fire fighter was pinned in a pile of burning wood, but extricated himself with great difficulty, and made his way out in flames which

---

[9] *Pollack on Torts,* 11th ed., Section 528.
[10] Smith v. Twin State Gas and Electric Company, 83 N.H. 439, 144 A. 57 (1928), at pp. 60 and 61.
[11] For a further discussion of this subject, see an article by James W. Morgan in *Fire Engineering,* Jan. 1964, p. 60.
[12] Dini v. Naiditch, 20 Ill.2d 406, 170 N.E.2d 881 (1960).
[13] Gibson v. Leonard, 143 Ill. 182 (1892).

he extinguished by jumping into a puddle of water at the curbing. He had to have 73 operations for skin grafts and for the reconstruction of eyelids and ears, and removal of scar tissue.

In reviewing the history of the labeling of fire fighters as licensees, the court said:

"It is our opinion that since the common-law rule labeling firemen as licensees is but an illogical anachronism, originating in a vastly different social order, and pock-marked by judicial refinements, it should not be perpetuated in the name of 'stare decisis.' That doctrine does not confine our courts to the 'calf path,' nor to any rule currently enjoying a numerical superiority of adherents. 'Stare decisis' ought not to be the excuse for decision where reason is lacking."

The court also held that the lower court should not have set aside verdicts given by the jury for judgment in favor of the injured fire fighter and the captain's widow, for the "defendants failed to provide fire doors or fire extinguishers, permitted the accumulation of trash and litter in the corridors, and had benzene stored in close proximity to the inadequately constructed wooden stairway where the fire was located . . ." and so the jury could have found that the defendants failed to keep the premises in a reasonably safe condition and that the hazard of fire, and loss of life fighting it, was reasonably foreseeable.

"Fireman's Rule"

The so-called "fireman's rule" that a land occupier is not liable for injuries to fire fighters from ordinary negligence, but only for wanton negligence, was discussed at length in Krauth v. Geller.[14] In this case, where a fire fighter was injured in a fall from the second story of a building under construction because a handrail had not yet been installed, the New Jersey Supreme Court sustained a finding of the appellate court that the land occupier was not liable. The fact that the fire department had responded to the building several times before due to an overheated salamander was not found to be negligence, per se, on the part of the owner or to provide adequate notice of the hazardous condition created by the lack of the handrail. Because of public policy, the court found that it would be too burdensome to hold a land occupier responsible for an injury to an "expert" who is paid to handle negligently caused fires. It pointed out that, by the very nature of their work, fire fighters are subject to hazards above and beyond those experienced by other workers, and that is why they are paid a good salary and receive extra pension benefits.

Here is another example: In ruling that a suit was deficient for failing to allege that the explosion which killed a fire fighter while fighting a fire was the result of an unusual, serious, hidden danger which could not have been anticipated, the Oregon Supreme Court stated:

"Cluttered conditions of the premises, and buildings constructed other than

---

[14] Krauth v. Geller, 31 N.J. 270 (1960).

in accordance with fire ordinances, which enhance the danger from fire are so usual as to be anticipated . . . oil in service stations and explosives in a munitions plant are also to be expected, because these substances, which are dangerous to fire fighters, are commonly identified with such places. . . . On the other hand, the unlawful storage of a fifty-gallon drum of gasoline in the basement[15] of a residence would be an example of a highly dangerous, hidden, and totally unexpected situation. If a fireman neither discovered nor learned of it, he would not assume the risk of injury of a resultant explosion."[16]

Following the same reasoning, the Court of Appeals of Washington affirmed a judgment of the county court denying recovery for the death of a fire fighter as the result of breathing heavy creosoted smoke at a pier fire. "The pier was afire when the decedent arrived at the scene. The circumstances of the fire were readily apparent to him. We cannot say that the defendant stevedore company had superior knowledge of the dangerous conditions. If anybody had superior knowledge of the dangerous conditions and risks involved, it was the decedent as a fireman . . . From the smoke's color and odor the creosote treatment of the pier would be obvious to a fireman . . . That creosote causes a more hazardous fire is also a matter within a fireman's expert knowledge."[17]

Although the "fireman's rule" would seem to exempt the land occupier from liability for injuries resulting to a fire fighter except in the case of wanton negligence, he may be held liable even where he is not negligent at all where the hazard encountered is so dangerous as to justify the imposition of absolute liability. Such was the case where a Louisiana fire fighter was hospitalized as a result of breathing antimony pentachloride gas which had escaped from the defendant's plant. The Supreme Court of Louisiana held that proof that the gas escaped is sufficient, and proof of lack of negligence and lack of imprudence will not exculpate the defendant corporation.[18]

Application of Statutory Safeguards to Fire Fighters

The courts are practically unanimous in holding that the violation of a statute or ordinance designed for the protection of human life or property is prima facie evidence of negligence. An injured party thereby has a cause of action, provided he comes within the scope of the particular law, and the injury has a direct and proximate connection with the violation. In determining whether the violation of safety ordinances gives an injured fire fighter a right of action against the guilty party, there is no unanimity in the case law. Apparently

---

[15] In Bandosz v. Daigger and Company, 255 Ill. App. 494, 141 A.L.R. 584, 592, a recovery was allowed in a wrongful death action where a fire fighter was killed by the explosion of flammable liquid stored in a basement in violation of an ordinance and statute.

[16] Morse, H. Newcomb, "Legal Insight: Building Owner's Duty vs. Fire Fighter's Safety," *Fire Command!*, Vol. 33, No. 3, March 1972, pp. 36, 37.

[17] Morse, H. Newcomb, "Legal Insight: Fire Fighter's Knowledge of Fire," *Fire Command!*, Vol. 38, No. 12, Dec. 1971, pp. 26, 27.

[18] Morse, H. Newcomb, "Legal Insight: Gassed by Antimony Pentachloride," *Fire Command!*, Vol. 40, No. 11, Nov. 1973, pp. 28, 29.

the crucial question in each case is whether it was intended that a fire fighter should come within the scope of the protection afforded by the statute or ordinance.

Where there are statutes or ordinances requiring safety devices or precautions, such as elevator shaft guards, some courts hold that the acts were passed for the benefit of all persons lawfully on the premises, including fire fighters and policemen. In Parker v. Barnard[19] the Massachusetts court, after holding the building owner liable to a policeman who fell through an unguarded elevator shaft, added, "Were the case at bar that of a fireman, who, for the purpose of saving property in the store, or for the performance of his duties, and who was injured because there was no railing and trap doors guarding the elevator well, he would have just ground for complaint that the protection which the statute had made it the duty of the owners or occupants to provide had not been afforded him."

In a similar New York case the court held that the board of fire commissioners, and not the injured fire fighter, was the proper party to invoke the statutory remedy for breach of law with respect to failure to keep hatchways and shafts closed at the end of business hours and the giving of damages for injuries sustained therefrom.[20]

Also, a Pennsylvania court permitted an injured fire fighter to recover from a landowner who violated a statute requiring elevator shafts to be kept closed, stating that the law was not restricted to a particular class but was general in its terms, and "it is a reasonable construction to hold that it was passed for the benefit of all persons lawfully on the premises."[21]

The Illinois court, in Dini v. Naiditch, discussed earlier in this chapter, held that a municipal fire ordinance which provides for the enclosing of stairwells, for the installation of fire doors and fire extinguishers, and for the regulation of waste disposal, and which then states that such provisions are intended "to prevent a disastrous fire or loss of life," is not limited in its operation to employees or tenants or to any particular class of persons. It was enacted, the court said, for the benefit of all persons lawfully on the premises, including fire fighters who are there to fight a fire, and evidence of violations of the ordinance is prima facie evidence of negligence.

A majority of jurisdictions, however, which have passed on the question have held that safety statutes do not apply to fire fighters and policemen. Typical of the reasoning offered by the court in cases denying liability to fire fighters under such safety statutes is Beehler v. Daniels,[22] where a Providence, R.I., fire fighter was injured because the occupant had stacked his merchandise so as to form a passageway directly to an open elevator shaft. In refuting the fire fighter's argument that the defendant should have anticipated the pos-

---

[19] Parker v. Barnard, 135 Mass. 116, 119; 46 Am. Rep. 450 (1883).
[20] Eekes v. Stetler, 98 App. Div. 76, 90 N.Y.S. 473 (1904).
[21] Drake v. Fenton, 237 Pa. 8, 85 A. 14 (1912).
[22] Beehler v. Daniels, 18 R.I. 563, 29 A. 6 (1894). See also Woodruff v. Bowen, 136 Ind. 431, 34 N.E. 1113 (1893), where eleven fire fighters fell four stories to their deaths when the building collapsed under the heavy floor load of water-soaked paper; also Litch v. White, 160 Cal. 497, 117 P. 515 (1911), where it was not shown that the owner had knowledge of the defective awning from which the fire fighters fell.

sibility of fire, the court maintained that if the argument were carried to its logical extreme, then every building owner would have to keep his premises well guarded at all times on the mere possibility that a fire fighter may enter who is unfamiliar with his place, and who might, in groping around in the smoke and darkness, stumble on some unguarded obstacle or opening. Recognizing the handicap of fire fighters in this respect, one court felt that the matter was worthy of legislative consideration.

Application of National Fire Protection Standards to Fire Fighters

Even if there is no ordinance specifying safety devices of adequate design to prevent a fire or explosion, a landowner or factory operator may be found liable on the basis of failing to adhere to nationally recognized standards, such as those promulgated by the National Fire Protection Association.

For example, in Bartels v. Continental Oil Company,[23] an action was brought for the wrongful death of a fire captain who died as a result of injuries suffered when a gasoline tank exploded while he was fighting a fire in the defendant's bulk storage plant and filling station.

A kerosene tank had already ruptured, providing additional fuel for the flames issuing from a tank truck that had been ignited by the truck's driver when he demonstrated his new cigarette lighter while loading his truck. Within the next half hour, two more storage tanks ruptured, and shortly after that the fourth tank left its concrete cradle and "rocketed" about 100 feet over the filling station; its 15,000 gallons of gasoline formed a ball of fire that engulfed several crews of fire fighters, killing one bystander and five fire fighters, and injuring twenty-three people.

Because the oil company's safety director was aware that these old storage tanks did not have vents of sufficient capacity, and the plant management knew full well that larger emergency vents would be safer, the Supreme Court of Missouri found that this hidden danger, though known to the defendant, was unknown to the fire fighters. Further, the oil company could have reasonably expected that such tank might rocket, yet it failed to warn the fire fighters of this risk.

The court acknowledged that under these conditions, an experienced fire captain would accept the presence of kerosene and gasoline as a known hazard of a fire in a gasoline storage facility. But, it declared, the law does not compel fire fighters to assume all possible lurking hazards and risks; it cannot be said that a fire fighter has no protective rights whatever. And it reaffirmed the exception to the general rule of nonliability on the part of a landowner to fire fighters, which imposes liability where the fire fighter is injured by a hidden danger on the premises, when the owner or occupant knew of the danger and had an opportunity to warn the fire fighters of it.

---

[23] Bartels v. Continental Oil Co., 384 S.W.2d 667 (Sup. Ct. Mo. 1964).

Status of Fire Fighters Called Outside of City Limits

Where fire fighters respond to a fire outside the city limits at the request of the owner, they are under no duty to be there, and are held to be invitees. Under such circumstances the owner must use reasonable care to see that his place is reasonably safe for the fire fighters he has invited to come there, and if there are dangers on the premises not obvious to the fire fighters of which he knows, or which with reasonable care he would have discovered, it is his duty to give warning of such dangers.

Fire fighters are expected, however, to know all the common hazards of fire fighting, and a failure to warn them of ordinary dangers encountered at fire, such as the possibility of a roof collapsing because of the added weight of a pipe line running over it, will not be considered negligence.[24] A Colorado court denied a fire fighter's widow damages where neither the occupant nor the fire fighters knew that the "smoke" which caused the alarm was deadly nitric acid fumes, resulting in the fire fighter's death.[25]

Public's Liability for Active Negligence

If the property owner has been guilty of active negligence toward the fire fighters, though the latter be classed as licensees, the owner is liable for the injuries caused. For example, where a fire chief was killed by an explosion of a burning box car loaded with fireworks, the Texas court held that the defendant should have anticipated that the rough handling of the box cars would lead to a series of explosions and that the fire fighters who would be called to fight the fires would be endangered. In granting a judgment, the court declared that the defendant was guilty of active negligence.[26]

Duty Owed by Utility Companies

Apart from the landowner-fire fighter relationship, another question arises respecting the duty owed by power and gas companies to safeguard their equipment so as to prevent injury to firemen. In a California case a Fresno fire fighter was electrocuted by a wire which had fallen at a fire.[27] The fire department had not requested to have the power turned off when the wire was discovered, but an agent of the company, who happened to be there as a spectator, knew that the line was down. The court held that there was no negligence in the power company's failure to have an employee on duty at the fire in the absence of an ordinance requiring one to be sent, or a request by the fire department. The court distinguished a Missouri case where a St. Louis fire-

---

[24] Buckeye Cotton Oil Co. v. Campagna, 146 Tenn. 389, 212 S.W. 646 (1922).
[25] Lunt v. Post Printing and Publishing Co., 48 Colo. 316, 110 P. 203 (1910).
[26] Houston Belt and Terminal Ry. v. O'Leary, 136 S.W. 601 (Tex. Civ. App. 1911).
[27] Pennebacker v. San Joaquin Light and Power Co., 158 Cal. 579, 112 P. 459, 31 L.R.A. (N.S.) 1099 (1910).

man was killed by stepping on a wire, on the ground that in the latter case the wires were in a public alley and in the California case they were on private premises.[28]

In reversing a judgment obtained in behalf of the fire fighter's widow, the California court denied liability on the part of either the landowner or the power company, but added the following warning: "We would not from this be understood as holding that in all cases where the owner is exonerated, an electric light or power company, using the building as a means of transmitting into or over it power in dangerous quantities, would also be exonerated." [29]

Where four firemen were electrocuted while turning the crank which lowered the aerial ladder, a Nebraska court held that there was no breach of any legal duty owed them by the power company. In denying the applicability of *res ipsa loquitur* the court said, "It is only when the defendant is under an absolute duty to prevent results (here, the electrocution) that their appearance shows negligence," and the court also found that this was not a case within the rule of absolute liability.[30]

Careless as the conduct may have seemed, a New Hampshire court did not find the fire chief's act of striking a match in a gas filled basement (to check a gas leak) such contributory negligence as would bar a recovery against the gas company. "But when the act is performed, as here, in the course of official service, if not as a legal duty, it is not to be said that performance of service or duty is negligence which will bar recovery."[31]

In a case Merritt v. Oklahoma Natrual Gas Co.[32] where serious injuries were sustained in the act of rescuing a person whose clothes caught on fire through the negligence of a gas company, the court declared: "The authorities are practically unanimous to the effect that one who imperils himself in order to rescue a person who is in danger of being injured or killed through the negligence of another person may recover damages from the negligent person for injuries received while effecting such rescue."

In Wagner v. International Railway Co.[33] it is stated by Justice Cardozo in part: "Danger invites rescue. The cry of distress is the summons to relief. The law does not ignore these reactions of the mind in tracing conduct to its consequences. It recognizes them as normal. It places their effects within the range of the natural and probable. The wrong that imperils life is a wrong to the imperiled victim; it is a wrong also to his rescuer. . . ."

In Norris v. Atlantic Coast Line Railroad Co.[34] it is stated: "This being true, it is well established that, when the life of a human being is suddenly subjected to imminent peril through another's negligence, either a comrade or a bystander may attempt to save it, and his conduct is not subjected to the same exacting rules which obtain under ordinary conditions; . . . It is always required in

---

[28] Gannon v. LaClede Gaslight Co., 145 Mo. 502, 46 S.W. 968, 47 S.W. 907 (1898).
[29] See Note 27, p. 463 of Pacific Report.
[30] New Omaha Thompson-Houston Electric Light Co. v. Anderson, 73 Neb. 84, 102 N.W. 89 (1905), at p. 96.
[31] Smith v. Twin State Gas and Electric Co., 83 N.H. 439, 144 A. 57 (1928), at p. 59.
[32] Merritt v. Oklahoma Natural Gas Co., 196 Okla. 379, 165 P.2d 342 (1946).
[33] Wagner v. International Ry., 232 N.Y. 176, 133 N.E. 437 (1921).
[34] Norris v. Atlantic Coast Line Railroad Co., 152 N.C. 505, 67 S.E. 1017 (1910).

order to establish responsibility on the part of defendant that the company should have been in fault, but, when this is established, the issue is then between the claimant and the company; and, when one sees his fellow man in such peril, he is not required to pause and calculate as to court decisions, nor recall the last statute as to the burden of proof, but he is allowed to follow the promptings of a generous nature and extend the help which occasion requires, and his efforts will not be imputed to him for wrong, according to some of the decisions, unless his conduct is rash to the degree of recklessness; and all of them say that full allowance must be made for the emergency presented. This principle is declared and sustained in many well-considered and authoritative decisions of the courts and in approved textwriters, and prevails without exception so far as we have examined."

Summary of Public's Duty Owed to Fire Fighters

1. **Duty of Owner-occupant**
    (a) To avoid wanton or wilful misconduct, and active negligence.
    (b) To warn of hidden dangers in the nature of a trap about which the possessor knows, but which would not ordinarily be encountered in that type of occupancy.
    (c) To maintain the ordinary means of access to the premises in a safe condition, as well as the fire escapes, smoke towers or other safety appliances required by law.
    (d) To allow fire fighters access to all portions of the building, whether to fight fire therein, or in adjoining property.
2. **Duty of Utility Companies**
    (a) To maintain their dangerous instrumentalities in a reasonably safe condition.
    (b) To repair dangerous conditions in their equipment which may lead to the injury of fire fighters, where:
        1. Notified by fire department to do so, or where
        2. Required by law to send a repairman to the scene of the fire or danger upon receiving official notice thereof.
3. **Duty of Motorists**
    By statute or ordinance motorists are generally forbidden, among other things, to follow closely behind fire trucks, drive over fire hose, etc., and must yield the right of way to fire apparatus giving the proper signal.

Review Activities

1. Describe how, in areas where there is no ordinance specifying safety devices of adequate design to prevent a fire or explosion, a landowner or factory operator may be found liable for injuries to fire fighters.
2. In your own words, write a definition of the so-called "Fireman's Rule."
3. Are fire fighters licensees? When classified as licensees, when can a property owner be found guilty of negligence?

4. Write a description of your state's regulations concerning statutory safeguards to fire fighters.
5. Explain how, if the property owner has been guilty of active negligence toward the fire fighters even though the fire fighters are classed as licensees, the owner is liable for the injuries caused. Include an example in your explanation.
6. Write an explanation of your state's policy concerning the duty of power and gas companies in regards to safeguarding their equipment so as to prevent injury to fire fighters.

Chapter Thirteen
LIABILITIES OF FIRE FIGHTERS

In General

The general rule, both by court decisions and by statute, is that every person is bound to conduct himself so that he causes no injury to the person or property of others. A person may be liable for injuries unintentionally inflicted, as well as those committed deliberately, where his conduct fails to come up to the standard of care exercised by the ordinary prudent man under like circumstances. There are certain occasions when a person may be liable for damage sustained by others even though he had done everything humanly possible to prevent the injuries; this rule of absolute liability extends, however, to those rather infrequent situations where a person engages in an activity so fraught with dangers to others that public safety demands that it be done at practically an insurer's risk.

But injuries for which the law allows redress are "legal" injuries, and not every omission which results in injury is cognizable in a court of law. To determine that a cause of action exists in the ordinary tort case, the following issues must be decided in the affirmative, and frequently by a jury: (1) Was there a duty owing to the injured party, or to the class of persons of which he was a member? (2) Was there a breach of this duty, i.e., was the duty wilfully or unintentionally omitted, or performed without the ordinary care exercised by a reasonably prudent man? (3) Was the breach of duty, if any, the proximate cause of the injury? (4) Was the conduct of the injured party such that his own acts did not contribute to his injury? In addition to determining the above issues, the jury must also decide what sum of money will most nearly compensate any detriment thus caused to the person or property.

While there is no unanimity of opinion on the question of whether fire fighters are public officers or mere employees, in those jurisdictions favoring the former view fire fighters are subject to the duties of care imposed upon them as individuals, and in addition, to those duties imposed upon them by reason of their status as public officers. But for failure to perform those duties imposed upon him by virtue of statutory or constitutional enactments, which are generally for benefit of the public as a whole, there is said to be no liability to any individual citizen, largely because he cannot ordinarily show that he has in fact suffered any pecuniary damage to himself as an individual over and above that sustained by all the members of the body politic. An illustration of the latter point is a situation where the city charter imposes the duty upon the fire department of enforcing the fire ordinances, and where that department is unable or unwilling to effectively carry out such a program. It must be remembered, however, that fire fighters may be subjected to criminal liability for failure to perform their duty.

Acting Within "Scope of Authority"

As for the liability of the fire fighters, themselves, it is generally held that they are not liable when performing acts—in good faith—which require the exercise of discretion or judgment, as long as they are within the scope of their authority. (For acts committed outside the scope of their authority, they will be liable in the same manner as any other citizen.) Courts are usually liberal in construing acts to be essential to one's duties, even to including collateral activities which, with proper motives, may tend to serve the basic purposes.

In California, as well as in the federal cases, the doctrine of official immunity has been expanded far beyond the limited degree to which it has been accepted in most states. The scope of public officers protected by the California doctrine today includes judges, health officers, policemen, city engineers, county clerks, and many more. The present law has been summarized generally as extending personal immunity not only to judicial and quasi-judicial personnel, but to all executive public officers when performing within the scope of their power acts which require the exercise of discretion or judgment. For torts committed outside the scope of authority, of course, personal liability would obtain as in the case of others who are not public employees. However, the concept of "scope of authority" for purposes of applying the immunity doctrine is exceedingly broad, and covers not only those duties essential to the performance of the job, for which the office exists, but also such collateral activities which, if engaged in with proper motives, would reasonably be construed as serving the basic purposes.[1]

Although the "scope of authority" may be broad, yet the act may not be considered discretionary, as in the case where a building inspector finds that a building meets all the code requirements, the issuance of a permit may be considered a wholly ministerial act, and the failure to issue it could result in personal liability to the city officials involved.[2]

It is generally conceded that the operation of government would be severely handicapped if officials whose very function and duty require the making of decisions involving judgment and discretion were to be held answerable in damages for mistakes, negligence or poor judgment in the honest performance of their duties. Fire fighters quite often have to make discretionary decisions, for example, whether to tear down a building in the path of a fire, or to destroy

---
[1] White v. Towers, 37 Cal.2d 727, 235 P.2d 209 (1951).
[2] Armstrong v. City of Belmont, 158 Cal. App.2d 641, 323 P.2d 999 (1st. Div. 1958).

a fence to gain access to the rear of a building to attack a fire, and their actions cannot be easily "second-guessed" by a group of well-meaning citizens sitting on a jury. "Obviously, public officials should not be exposed to risks of this magnitude. The policy behind the immunity doctrine—to promote fearless performance of duty—as well as the practical impossibility of drawing any rational dividing line between discretionary and ministerial acts, strongly argues that personal immunity should attend all public officers and employees in the good faith performance of acts within the scope of their authority."[3]

Liability as a Public Officer

There are situations where the public officer's delegated power is strictly construed, and if a fire fighter, purporting to act as an officer of the city, causes legal detriment to an individual, there may be civil responsibility in damages:

1. **No Power to Act**

Where the power to act in a given field does not exist at all. For example, if the authority given by law is to extinguish and prevent fires, he has no authority to tell a man how to construct a building.

2. **Beyond Authority to Act**

Where the power exists but the fire fighter acts beyond the scope of his authority. For example, the fire fighter's duties may be limited by law to the extinguishing and prevention of fire, and in case of extreme emergency to perform tasks involving the saving of life and property. However, the court would be likely to give a liberal interpretation of such powers in the event of a bombing attack, earthquake, flood, or similar catastrophe, but might take a narrower view otherwise.

Examples of situations where liability could be imposed for acting outside the scope of authority include the following:

Riot Suppression

Where the authority of the fire department is limited to extinguishing and preventing dangerous fires, and enforcing fire laws, fire fighters cannot lawfully use their hose streams and wagon batteries to assist the police in quelling riots; such action would be clearly *ultra vires* and render the fire fighters personally liable for any injuries caused by such tactics. However, if the fire fighters were engaged in fighting a fire and were attacked by rioters, any reasonable measures adopted by them in self defense might be justified, but in no case would this extend to the use of firearms or the use of excessive force. Under such circumstances the fire fighters should call for law enforcement officers to protect them, and if such help is unavailable, they should withdraw to a position where they would not be required to inflict injury on others as a means of merely performing fire suppression activities.

If their position must be maintained to perform rescue work, then only such force as is necessary to accomplish such objectives would be justified.

---
[3] 10 U.C.L.A. Law Rev. 463 (1963), at p. 483.

Rescue Work

While there should be no doubt that the saving of human life as an incident to fire extinguishing is clearly within the province of a fire fighter's duties, yet when he responds to hospitals or private homes merely to operate resuscitators or other respiratory apparatus, he is probably acting in an *ultra vires* capacity unless there is legal authority for rendering such services.

His humanitarian motives may not act as a bar to liability if, through failure to use due care, he injures the subject of his oxygen therapy. Of course, if the fire department has charter authorization to operate an ambulance service and perform general lifesaving activities, then the rescue of drowning victims would be *intra vires*.

A Seattle, Washington, lawsuit involving two fire fighters who were responding to a child locked in a bathroom raised the question as to whether they were acting within the scope of their authority and hence whether they had the right to use their red light and siren.

In Illinois, the statutes dealing with fire departments refer to fire extinguishing and fire prevention but make no reference to rescue work as part of the purpose for which they were created.

In Los Angeles it took a charter amendment and a special ordinance to grant the fire department specific authority to save lives and property (in addition to extinguishing and preventing fires). The California Government Code (Sec. 850.8) covers this subject now. (See Appendix C.)

About three-fourths of the states have "Good Samaritan Laws." A Good Samaritan Law provides that one who voluntarily assumes to help someone who has been injured shall not be chargeable with any fault or be responsible at law for any errors or omissions in the care that he renders. It provides an immunity though not an absolute one. Almost every state has different language in its statutes. Most of them provide that the exception to the rule is gross negligence or wilful or wanton misconduct which results in some injury to the individual, in which case, liability will result.

Some states have extended the "Good Samaritan" coverage to paramedics. For example, California state law provides that a paramedic who follows the instructions of a doctor or nurse will be immune from civil liability to the same extent that physicians and nurses are now protected when rendering voluntary emergency services.

Salvage Work

Under the power to extinguish fires and save lives, fire fighters responding to an alarm of fire need have no apprehension about activities designed to prevent water damage to the property involved, whether the water issues from hose streams or automatic sprinkler systems. But where the call is in no way related to a fire, as where fire fighters are called to pump out a cellar flooded by a broken pipe, or to patch up a homeowner's roof in a rain storm, then here again their endeavor takes them beyond their legal authority to prevent water damage resulting from fire fighting operations.

Demolition Squads

Some fire departments have a group of individuals who are trained in the handling of high explosives, so that if it becomes necessary to blast fire breaks in order to stop a conflagration, no time will be lost in carrying out such a plan. The power to dynamite buildings in advance of a fire is no longer questioned, but should a person be injured as a result of the negligent use of explosives while the fire fighters are doing such things as assisting road maintenance men in removing old bridge footings from a street, then liability cannot be averted on the ground that such acts were done in the course of duty.

3. Lack of Care in Exercising Power

Where the power exists and the proper method of exercising it has been followed, but with a lack of due care, thereby resulting in injury to an individual or to property which would not otherwise have occurred, liability can result. For example, a fire fighter who needlessly throws clocks and table lamps, or any other valued possessions of a householder out of a second story window in an effort to save the owner's goods from fire may be considered as failing to use reasonable care in the matter.

First Aid Volunteers

One who voluntarily assumes care of an injured person is charged with the duty of common and ordinary humanity, meaning that the individual should do the utmost in order to provide proper care and attention, and certainly must not attempt to make things worse.

In a case where a fire department rescue team tried to resuscitate a college student who had been forced to swallow a chunk of liver as part of a fraternity initiation, the air passage remained blocked and the boy died. The city was sued because of the alleged negligence on the part of the fire fighters in not transporting the victim to the receiving hospital for an emergency operation; damages were paid to the family of the victim.

A Federal District Court in Chicago ruled in 1975 that the fire chief of a rural village could be held liable to pay damages because a pregnant woman was refused ambulance service, if it was the policy of the fire department to provide service to nonpaying persons in life and death situations and the chief intentionally withheld services under circumstances where he would normally have rendered care. The village had discontinued free ambulance service in 1972, and normally only served residents in the unincorporated area who contracted for such service, except in life and death situations.

In this case, a multimillion dollar law suit was instituted because the woman's unborn infant died in spite of a Caesarean section when she finally got to the hospital. The court held that if the chief refused to respond with the ambulance it violated the woman's due process and equal protection rights, and referred the case to a jury to determine whether the chief was aware that the request was a life and death situation, and if he was, the extent of the damages.[4]

---

[4] FIRE DEPT. PERS. REP., No. 9, Sept. 1975, pp. 4–5.

Standard of Care in Rescue Service

In the absence of immunity statutes, and excluding wilful or wanton misconduct, a fire fighter can be held liable for negligent acts. (Negligence is the failure to do something which an ordinary prudent rescue fire fighter would have done, or the doing of something which the ordinary prudent rescue fire fighter would not have done.) As a general rule, the fire fighter's duty is to carry out those rescue efforts for which he is trained, and in the manner in which he has been trained. Practically all fire fighters today are trained in standard first aid practices, and all rescue personnel are properly instructed in closed chest-heart resuscitation. If a situation should arise where this procedure was called for in the presence of rescue personnel, a failure to at least attempt it might be considered negligence. But negligence is not limited to the failure to perform acts called for by cardio-pulmonary resuscitation (CPR) training; it can also be based on carrying out the required techniques in an improper manner. For example, Dr. William H. L. Dornette has cautioned that "pushing sideways instead of straight down, when performing closed-chest resuscitation, may fracture ribs and puncture the lungs. This misapplication of the technique can well be construed to be a negligent act, even though it is possible to break ribs if the technique is applied properly." After pointing out that a fire fighter does not have to possess the skill of a physician to attempt resuscitation, Dr. Dornette says that he would only be liable if he did not possess the degree of skill and care possessed by comparably trained and equipped rescue personnel. Moreover, the emergency nature of the situation is always taken into consideration in assessing the standard of care.[5]

Where a patient is unconscious, consent to administer rescue service may be implied from the emergency circumstances as being necessary to save his life; otherwise it could be considered a wrongful act to treat him against his will.

In the case of a bona fide emergency, treatment can be administered to a person who is unconscious or incapable of giving consent, even if his parent, spouse, or other relative refuses to consent on religious or other grounds. "While he must proceed with great caution, as he should anytime he treats a minor," Dr. Dornette says, "the rescuer should nevertheless proceed. In comparable cases involving blood transfusions, the courts have unanimously held that the refused treatment should be given, and that doing so in the absence of parental consent is not actionable."[6]

Care Required in Physical Rescues

Not everyone is grateful for being rescued. For example, in 1974 a man who was successfully rescued from his burning home where he had been trapped, later sued for damages due to the rough treatment he suffered while being saved. In this case,[7] the court construed the California Government Code

---

[5] Dornette, William H. L., "Fire Fighters, Resuscitation and the Law," *Fire Command!*, Vol. 38, No. 7, July 1971, pp. 13–15.
[6] *Ibid.*, p. 15.
[7] Heimberger v. City of Fairfield, 117 Cal. Rptr. 482 (Cal. App. 1974); FIRE DEPT. PERS. REP., No. 1, Jan. 1975, p. 9.

Section 850.4 (discussed in Appendix C) in a liberal way, so that the tort immunity granted fire fighters for injuries caused during fire fighting functions was held to include rescue efforts while fighting a fire.

Though an off duty New York City fire fighter was awarded a hero's medal for smashing through a hotel wall to rescue a couple trapped in an elevator, he was later convicted in a criminal action for having been belligerent to the policeman who arrested him following the rescue, and for damaging the hotel's elevator. On appeal the conviction was overturned and he sued the city and the officer for false arrest. After paying out $12,000 in legal fees over a period in excess of six years, the civil court jury awarded him a judgment for $30,000 damages.

4. When Acting Out of Private Motives

Where the power has been exercised by a fire fighter for a nonpublic purpose, to someone's damage, he is liable therefore; as for example where the motive in sounding the siren and pre-empting the right of way was not to respond to a fire but to take a friend to the airport.

5. When Acting Recklessly

Where a fire fighter uses his power wilfully or recklessly to the detriment of others, e.g., where he is driving a fire truck under circumstances permitting him to disregard the ordinary traffic laws, but in reckless disregard for the safety of other persons on the highway, he runs down a vehicle which had failed to heed his siren, then he cannot claim immunity from liability. (Driving liability is discussed in Chapter 6.)

An example of how this principle has been written into state law can be found in the California Government Code Section 850.8 relating to the problem of transporting persons injured at a fire. The code says that a fire fighter or other public employee, when acting in the scope of his employment, may transport or arrange for the transportation of any person injured by a fire, or by a fire protection operation, to a physician or hospital if the injured person does not object. Further, the public employee will not be held liable for any injury sustained by the injured person as a result of such transportation or for any medical, ambulance, or hospital bills incurred by or in behalf of the injured person or for any other damages, unless the injury was caused by his wilful misconduct in transporting the injured person or arranging for such transportation.

Thus, whenever a fire fighter departs from the terms of the power that has been granted to him as a public officer, he becomes just as responsible for his acts as if he had never been authorized to act in a public capacity.

Discretionary Powers and Ministerial Duties

Another matter which deserves consideration is the distinction between the nonliability for the failure to exercise a discretionary power and the possible liability for refusal to perform mere ministerial duties (duties which do not require the exercise of discretion). Where the power exists, such as that usually granted the chief (or his representative) to investigate fire hazards, but the time, place and mode of its exercise rests in the discretion of the officer, there is no civil responsibility to citizens for the exercise or non-exercise of this power.

But on the other hand, with respect to ministerial duties as, for example, the commonly prescribed duty of fire department boards or officers to issue certain permits upon the finding that certain facts exist, here the power, the time, place, and the mode of exercise are all prescribed by the law and the officer has nothing more to do than determine the existence of facts upon which he must act. If he refuses to act—one way or another—a writ of mandamus will issue to compel action of some sort where there is a positive duty to exercise the power.

Therefore, officers are liable for nonfeasance or negligent performance of a duty that is clearly set forth, where the means and ability to perform it are present and when its performance or nonperformance or the manner of its performance involves no question of discretion. "In short, where the duty is plain and certain, if it be negligently performed, or not performed at all, the officer is liable at the suit of a private individual especially injured thereby."[8]

Liability of Superior Officers

Under the doctrine of *respondeat superior* the principal is liable for the torts of the agent or servant committed while acting within the scope of his employment, even though he acts contrary to his instructions, and this liability extends to the wilful and malicious torts of an agent, as well as to negligent acts, unless, though in the employer's services, they are committed for the employee's own purposes. In the absence of *respondeat superior* there is no imputation of liability to the superior for the act of a subordinate officer. However, where he participates in, or directs the agent or servant to perform tortious acts, or is individually negligent, he is liable.

In California the doctrine of *respondeat superior* in the tort liability of municipal officers between themselves is not recognized. Hence, as between two officers under the same sovereignty, there is no master-servant relationship though one officer is subordinate to another. In a Massachusetts case a member of the Somerville fire department was driving the chief to a fire, but at the time he lost control of the car (through excessive speed) there was no evidence that he was acting under the personal direction or orders of the chief, nor that he was his agent or employee; hence, in an action by the injured person against the driver and the chief, the court found that the doctrine of respondeat superior was inapplicable. A judgment was rendered against the driver, and in directing a verdict for the chief the court explained that as public officers of the city the men were not fellow servants.[9] However, "ratification may be equivalent to command and cooperation may be inferred from acquiescence where there is a duty to restrain," said the New York court where a

---
[8] Doeg v. Cook, 126 Cal. 213, 58 P. 707, 77 A.S.R. 171 (1899).
[9] Sherry v. Rich, 228 Mass. 462, 117 N.E. 824 (1917).

fire commissioner failed to keep his driver from speeding, and an accident occurred.[10]

Therefore, the chief of a civil service conducted fire department, in appointing a subordinate, is not liable for the latter's torts, for both are "servants of the law" and there is no master-servant relationship between them. The liability of the chief arises only when he has directed, participated in, or ratified the tortious act or omission, so as to make it his own in legal contemplation. Although an officer, under ordinary circumstances, is liable for the acts of his deputy, who is his alter ego in contemplation of law, this is not followed where the deputy is under civil service, for here the officer has no choice in the selection of his deputies, and therefore should not be personally liable for his wrongs.

Need for Insurance

Fire fighters who drive public vehicles should obtain broad form insurance to protect them in case the injuries which they inflict upon others are found to have occurred while they are acting outside the scope of their employment. For example, if a city vehicle is driven upon a supermarket parking lot by a fire fighter who is shopping for a personal item, and in the course of parking the vehicle he negligently damages another car, the city attorney is likely to rule that this action was outside the scope of his authority, and will refuse to defend the fire fighter in a lawsuit. He may even require him to pay for the damage to the city vehicle.

Most insurance policies are not operative while the insured is driving a government vehicle, nor while he is driving his own vehicle on his employer's business. For example, if a fire fighter leaves the engine house and uses his own car (or a department vehicle) to go to the store to buy something for the fire house, he is acting in line of duty, but if he leaves and shops for himself, whether using his own or a fire car, he is acting outside the scope of his authority; in either case he should have broad form insurance in order to be covered in case of an accident.

Sample Problems

Having reviewed the laws that are applicable to a fire department, let us take a look at a specific example of a problem that commonly arises in the fire service, i.e., the possibility of liability for permitting persons to ride in fire department vehicles or for allowing persons to engage in athletic activities at a fire station.

From time to time requests are received from persons who are not members of the fire service for permission to play on department maintained handball courts, etc., to temporarily reside in a fire station, to ride to fires on apparatus, or to participate in training programs at drill towers. It has been the practice to require guests to sign a "visitor's waiver" whereby they purport to waive all claims of liability against the city, the fire department, or any individual member for personal injuries sustained while engaged in such activity.

---

[10] Dowler v. Johnson, 225 N.Y. 39, 121 N.E. 487 (1918).

It has long been recognized that such waivers have no validity whatever, for anyone who is injured through the negligence of another is not barred from obtaining a judgment for damages resulting from such negligence. Since most persons who sign waivers do not know that liability could be imposed anyway, there is less likelihood of having lawsuits instigated where such waivers have been obtained in advance. For this reason it is desirable, at least in the view of municipal attorneys, to require persons who are not members of the fire department to sign such waivers prior to engaging in any department activity involving an unusual risk of injury. It probably would not be necessary, for example, to require a waiver from a visitor who only plays a game of handball very infrequently; however, where outsiders wish to regularly participate in athletic activities on department facilities, they could be required to sign waivers and to use the protective safety items prescribed for regular members, such as the eye guards for the handball players.

Insofar as liability of the fire fighter is concerned, the following general principles apply:

Anyone can be held liable for injuries sustained by another person:

1. where he deliberately caused it to happen, or
2. where he negligently caused the injuries and the injured party was not contributorily negligent; even contributory negligence may not be a bar to liability in some states where comparative negligence is recognized.

Therefore, a captain who permits a visitor to play handball at his fire station will not generally be held liable for any injury to the visitor unless he deliberately does something which causes the injury, or he is actively negligent toward him and the visitor has in no way contributed to his own injury.

Negligence—What Is It?

"Negligence" is the failure to exercise the reasonable care that an ordinary prudent man would do under like circumstances. For example, an ordinary prudent man doesn't close the door to a handball court while another man's hand is in the opening. So if a fire fighter shuts the door on a visitor's hand, this would be negligence, for even if he didn't see the hand (because he failed to look) a prudent man would have looked first! Likewise, a prudent man doesn't stand out of sight with his hand in a door opening, because he should anticipate that some careless person may not see him, and may slam the door shut. So if his hand gets mashed in this manner, he could be considered contributorily negligent and in many states contributory negligence is a bar to recovery in a civil action. However, in those twenty-seven states which have adopted the doctrine of comparative negligence, damages might be reduced on the basis of the amount his own negligence contributed to his injury. Under the "pure" form of comparative negligence, such as the California Supreme Court enunciated in 1975, a jury could find that the door-slammer was 70 percent at fault and the visitor 30 percent at fault, and award him 70 percent of the damages sought by him for his injury.

In the nineteen additional states which have the "Wisconsin system" or "50 percent system," the plaintiff cannot collect damages if he is equally or more at fault than the person he is suing. But this would not necessarily discourage him from suing, regardless of where the injury occurs, or the percent of his carelessness.

Remember, anyone can sue anyone! But "suing" is not the same as "collecting." To "collect," a person must have a good cause of action, and the person sued must have no defense whatever. Oftentimes injuries occur where no one is at fault or both parties are at fault, and even though both parties are insured, the insurance companies will not "pay off" in these situations, unless, where both are at fault, the doctrine of comparative negligence can be applied.

The important thing is to warn any visitors entering fire department property of any hidden dangers such as holes, overhead objects which might fall, etc., and then refrain from any act which may cause injuries; naturally, greater care is usually required in safeguarding small children than adults.

As far as the typical handball injury is concerned, such as a broken toe or finger, sprained ankle, etc., these are risks assumed by the player, for which no legal redress is available. However, if the floor suddenly caves in, or the light fixture falls on him, there could be municipal liability if the injured party were able to establish that these dangerous conditions had been reported such a long time back that ample time had elapsed to have permitted their correction, and nothing had been done about it.

The fact that the man had signed a waiver form would not bar his recovery of damages for any injury sustained by these conditions; however, if he had been told that the fixture was loose and might fall, or that the floor was weak from dry rot, and he said he would take a chance and play on the court anyway, then his prospects of obtaining a judgment would be very slight because his playing with full knowledge of the danger could be considered contributory negligence or assumption of risk.

Under the above circumstances, the station commander, having reported the dangerous conditions to his superiors, would not be responsible for the injuries, since the correction of such conditions are not within his control, and if the conditions were not known by anyone prior to the time of the injury, and if reasonable inspections would not have disclosed the dangerous condition, then no one, including the municipality, would be liable for the injuries.

Reasonable Care—What Is It?

In the last analysis the main thing to remember is this: to avoid liability for damages to other persons, exercise the same reasonable care in all of your actions that any ordinary prudent man would do under like circumstances; this is the test of negligence. This is the test that your fellowmen, sitting as a jury, would apply to your conduct if you were brought into court in a lawsuit in a negligence case. It is also the test applied by a commanding officer in passing upon the reasonableness of any course of action his subordinates may take which is not otherwise expressly set forth in law or in department regulations.

Thus, while we have been talking mostly about the responsibility for injuries suffered by handball players who are not members of the department, the same principles of negligence and liability apply to any department facility and to any third party who participates in department activities, with the exception of riding in city vehicles. In the latter situation, a law usually covers the subject quite explicitly, the context of which is generally required to be set forth in some manner in every municipally-owned truck and automobile. In Los Angeles, for example, it reads as follows: "(b) . . . It shall be unlawful for the operator of any of the foregoing equipment (automotive equipment, automobile, truck or other motor vehicle owned by the City of Los Angeles) to permit, suffer or allow any person to ride in or upon the same unless the riding in or upon the same by such person is necessary for the execution of official business of the City of Los Angeles."

By way of example, the city attorney in Los Angeles has ruled that securing news stories by a newspaper reporter cannot be held to be official business and gave as his opinion that permission in such cases should be denied; he declined to answer what the effect of obtaining a waiver would be since it was immaterial, the action in itself being unlawful. Nevertheless, most fire departments will continue to obtain written waivers from all nonmembers who are given special written permission to ride on fire apparatus, under those circumstances where it can be done so lawfully (on official business). Even though such waivers may not have any validity, they are obtained in the belief that they may discourage lawsuits.

Strikes—What Are They?

A "strike" has been defined as the cessation of workmen as a means of enforcing compliance with a demand made on their employer.[11] This definition may imply a complete severance of the relations between the employees and the employer, and was held to be exactly that by a California Appellate Court on Nov. 25, 1975, when applied to public employees.[12] "'The right to strike' contemplates a lawful strike. There is no legislative enactment by the California Legislature which expressly authorizes strikes by public employees . . . Suffice it to say that public employees do not have the right to strike in this state." Strikes have been declared illegal by the highest courts of California through a succession of decisions dating back to 1949. The court declared that when the court clerks (petitioners) undertook to withdraw their services by engaging in a strike they effectively terminated their employment, and the employer was at liberty to hire others to perform the services.

Having exercised their constitutional right to quit work, they were no longer under their employer's control, and could not be ordered to perform duties incident to their former employment, and could not be found in contempt (and jailed) for refusing to work. One judge concurred, but on separate grounds, and one dissented, saying that, as officers of the court, they could be punished for

---

[11] Ballentine, James A., *Law Dictionary with Pronunciations,* Lawyers Co-Operative Pub. Co., Rochester, New York, 1930.
[12] *In re* Webb, 52 Cal. App.3d 648, at 649, Cal. Official Reports No. 32 (1975).

creating the chaos caused by their "withdrawing their services in concert."

Is There a Right to Strike?

A federal appeals court in Washington, D.C., in a 1971 ruling that could be interpreted to include all public employees, whether federal, state, or municipal, declared that public employees do not have the right to strike. It rejected the postal union's challenge to the constitutionality of the federal laws barring federal workers from striking.[13]

Strikes by fire fighters, policemen, teachers, sanitary workers, and other public employees have been declared illegal by local and state decrees in many jurisdictions. A California appeals court ruled in 1970 that municipal employees do not have the right to strike, and declared that no legal strike is possible without the express granting of that right by the legislature.[14]

This 1970 ruling by Justice Martin Coughlin of the 4th District Court of Appeal made possible the mandatory issuance of injunctions by superior court judges against any striking city employees in the state. He said: "The legitimate and compelling state interest accomplished and promoted by the law denying public employees the right to strike is not solely the need for a particular government service (in this case, sanitation workers of San Diego, Ed.) but the preservation of the system of government." He then quoted a statement of Franklin D. Roosevelt, in which the former president had written: "Particularly, I want to emphasize my conviction that militant tactics have no place in the functions of any organizations of government employees . . . a strike of public employees manifests nothing less than an intent on their part to prevent the operation of government until their demands are satisfied. Such action, looking toward the paralysis of government by those who have sworn to support it, is unthinkable and intolerable."[15]

What is the Liability for Striking?

What is the liability for illegal strikes? Where the court has issued an injunction against striking fire fighters, those who have refused to go back to work have been held in contempt in some jurisdictions. Where San Diego fire fighters continued to man picket lines despite a court order to the contrary, contempt of court actions against them were dropped as a term for settlement of the strike with the city. However, under the 1975 court decision, discussed earlier, striking fire fighters will hereafter be deemed to have abandoned their positions, and can be replaced, by the city with new recruits without delay.

---

[13] United Federation of Postal Clerks v. Blout, 325 F. Supp. 879, aff'd, 404 U.S. 802, 92 S. Ct. 80 (1971).

[14] California State Council of Carpenters v. Superior Court, 11 Cal. App. 3rd 144, 89 Cal. Rptr. 625 (4th Dist. 1970).

[15] The author concurs in this view, particularly with reference to those fire fighters who, until they commenced going out on strikes, had the high esteem and trust of the citizens whose lives and property they swore to protect. It also seems to him that there is public support for this view illustrated by the 1975 elections in San Francisco. At that time voters apparently retaliated against an illegal strike of fire fighters and police officers by the overwhelming adoption of three ballot propositions which adversely affected future benefits and working conditions in the fire and police departments. To the author, the San Francisco vote reflected more than anger with public employees who walk out in defiance of the law; it also reflected frustration over the total absence of laws that might prevent such strikes (there were over fifty involving public employees in 1975 in California alone) by imposing binding procedures for negotiation.

Pending the hiring of new replacements, the attorney general in California has ruled that state forestry fire fighters can be ordered into an area where fire fighters have gone on strike. His opinion concluded with, "Strike-hit communities can legally call in neighboring community police and fire fighting forces, in addition to state law enforcement and fire agencies."[16]

In addition to the possibility of dismissal or being jailed for contempt of court, illegal strikers may be subject to other penalties. In a Fayette County (Kentucky) Circuit Court decision,[17] where, despite a court restraining order, a majority of the Lexington fire fighters staged a ten-day strike to try to get their union recognized as their bargaining agent, the International Association of Fire Fighters Local was fined $10,000 and the union field representative was fined $500 for contempt of court. The judge also ruled that the more than 300 fire fighters would be subject to ten-day sentences if they again violated his orders during a one-year probationary period, and the union field representative would be subject to a thirty-day sentence during the same probationary period if he violated his orders.

Following the above court decision, the Lexington fire chief preferred charges against striking members who used fire apparatus to block an intersection at the beginning of the strike, seeking to have the captain's and lieutenants among them reduced in rank to fire fighter, and to have participating fire fighters reduced one step in grade and pay, because their actions constituted "unlawful and insubordinate control over vehicles" and was a serious breach of discipline and chain of command.

Records kept by a fire department of its officers and members who report for duty during a strike, which has averaged about five to ten percent of the force in large cities, can enable the city to pay the salaries of those reporting for work and dock the pay of strikers. In the Milwaukee strike, the fire fighters were paid during the strike because there were no records to show who was off duty, and in the final determination, the court refused to take away salaries already paid.[18]

In New York City, where only 10 percent of the 375 companies were in service at the start of a strike, the records kept provided the basis for a later action by the city that resulted in a $650,000 fine of the fire fighter's union.

Picketing—Is it Legal?

An appellate court of Illinois, after holding in 1968 that public employees have no right to strike and granting the City of Rockford an injunction against a strike by its fire fighters, declared that picketing is a valid method of expression and ordinarily an injunction will not be granted against peaceful picketing for a lawful purpose. However, if it could be shown that picketing of the city hall and fire station actually interfered with governmental functions, it could be enjoined.[19]

---

[16] Opinion of Attorney General Thomas Lynch, reported in *League News*, Feb. 1971, p. 2.
[17] Lexington-Fayette Co. v. Lass *et al*, Fayette Co. Cir. Ct. (Ky. 1974); FIRE DEPT. PERS. REP. (sample issue), pp. 13–14.
[18] *Fire Engineering*, Aug. 1974, p. 90.
[19] City of Rockford v. Local No. 413, Int'l. Assn. of Fire Fighters, 240 N.E.2d 705, App. Ct. of Ill. (1968).

Strike Alternatives

Where strikes are illegal, there is no validity in depriving the public of the fire protection the fire fighters are paid to furnish by staging "slowdowns" or "sick-outs." Such methods of attempting to coerce the political entity to grant the objectives of fire fighters are just as illegal and just as morally reprehensible as a strike.

Most fire fighters would rather seek their legitimate aims through legal procedures. Where state laws have been adopted relating to public employee organizations, they have the right to expect the management representatives of their political subdivision to comply with the law and bargain in good faith. However, when the governor has the power to order state fire fighters to be placed in municipal fire stations, there is not likely to be motivation to seriously bargain with fire fighter's representatives that might otherwise exist, knowing that fire fighters are not likely to abandon their positions, or threaten to do so, if adequate fire protection can be quickly furnished in their absence.

It would appear that the most reasonable alternative to strikes would consist of state or federal collective bargaining legislation for public employees which will at a minimum provide for (1) a method of impasse resolution through compulsory and binding arbitration with costs borne equally by both parties; and (2) that public safety employees shall have the right to organize and bargain collectively in bargaining units consisting at a minimum of all members in existing fire department organizations. However, compulsory arbitration statutes have not met with universal approval in the courts. On October 9, 1975, the Supreme Court of South Dakota held that a statute providing for the arbitration of impasses and grievances was an unconstitutional delegation of legislative powers.[20] The opposite view was reached by the highest court in New York, on appeal from decisions affecting the cities of Amsterdam and Buffalo.[21]

Review Activities

1. Write an explanation of "legal" injuries. Include in your explanation the issues that are determined in a legal injury case.
2. Define "negligence," as related to rescue fire fighters.
3. Describe the rules that govern athletic activities at fire stations in your community.
4. With a group of your classmates, discuss "scope of authority." Then, consider the following circumstances in relation to a fire fighter's scope of authority. In which of these circumstances would a fire fighter be liable, and in which would a fire fighter not be liable? Defend your reasoning by explaining why.
    (a) A hurricane.
    (b) A riot.

---

[20] City of Sioux Falls v. Sioux Falls Fire Fighters, 234 N.W.2d 35 (S.D. 1975); FIRE DEPT. PERS. REP., No. 11, Nov. 1975, p. 3.

[21] City of Amsterdam v. Helsby, and City of Buffalo v. New York State Employment Relations Board, 37 N.Y.S.2d 19, 371 N.Y.S.2d 404, 332 N.E.2d 290, (1975) reversing 362 N.Y.S.2d 968 and affirming 363 N.Y.S.2d 896.

(c) A child stranded in a tree.
(d) An elderly woman who has stopped breathing.
5. Does your state have a "Good Samaritan Law"? Describe such a law, including in your description reasons why volunteer first aid care can be dangerous.
6. In a general class discussion, consider your community's laws, if any, regarding a fire fighter's right to strike. Elicit discussion and answers concerning each of the following:
    (a) Do you feel fire fighters have the right to strike? Why, or why not?
    (b) Examine a past or present fire fighter's strike (from any area in the country). What were the fire fighter's reasons for the strike? How was the strike settled? Do you feel that striking was an appropriate means for the fire fighters to make their point? Explain.
    (c) What are some alternatives to striking in order to rectify a situation?

Chapter Fourteen
PROCEDURAL POINTERS

Surveys in the Field

Fire inspectors will find that many people resent the idea that they must be told in writing to do some simple thing which they will readily agree to do right away. If the matter is one which can be corrected while the inspectors are still on the premises, it is good practice to encourage this to be done. However, if the item requires a little time because of its magnitude, the manner in which the recommendations are presented will have an important influence on the recipient's reaction; the inspector should, in the case of a lengthy survey, prepare a statement of just what the conditions were that he found upon the premises, supplementing it with drawings or sketches (photographs where necessary). The second part of the survey should be made up of the specific things which must be done to correct the condition. The list should be in such form that an owner or manager can take the list to an architect, sprinkler contractor, or a builder and they will be able to understand exactly what is to be done.

Written Notices

Where the conditions do not warrant an elaborate survey, but yet the hazards are such that it is necessary to keep a record of having advised the parties involved of what to do, fire inspectors frequently write "notices" to the occupant of the premises, outlining concisely the fire protection standards which must be met. Sometimes these notices are printed in the form of a "recommendation memorandum," thus avoiding the impression of giving a man a "ticket" or ultimatum on the first visit. If, on subsequent reinspection, it is found that there has been no compliance with these recommendations, and there appears to be no justification for the oversight, then it is customary to give the occupant or owner or both a "final notice." If this is not obeyed in the stipulated time and the hazard is not one that is of such a serious nature as to require immediate action, the fire department procedure may call for giving a formal notice to the violator directing him to appear in the office of a deputy city attorney at a given time, wherein he is asked if he has any good reason why a complaint should not be filed against him. Oftentimes this admonition and hearing is sufficient to clear up any misunderstanding that may have arisen.

Thomas Williams, a former Deputy City Attorney in Los Angeles, has described some of the fundamentals involved in obtaining enforcement of fire ordinances. The fundamentals, summarized in succeeding paragraphs, are based to some extent upon Mr. William's descriptions, augmented by the author's own experiences and from other sources.

The Inspector's Demeanor

The manner in which the inspector conducts himself in the field has an important bearing on the successful outcome of such procedure. He should realize that he often is the initial contact between the municipality and the public, and the establishing of good public relations will depend a great deal upon his demeanor. He should demand respect but at the same time keep in mind that the "greater the man the greater the courtesy"; the amount of respect that he has a right to demand depends upon how well he knows his fire prevention principles and ordinances, how neat he is in appearance, how well he handles his work with authority, tact, consideration for the other man's problems and his willingness to devote the necessary time to find all the facts, research the problems and arrive at definite recommendations. He should be impersonal—never show any visible agitation over petty matters—and in writing his notices, specify exactly what he wants done. The notices should be complete; he should not just put one or two items on the original recommendation memorandum and then when it becomes necessary to cite the man into the city attorney's office put a dozen items on the "white sheet" (letter of citation).

The inspector should be careful to give the notice to the person who is responsible for the violation. In court he may be called upon to identify the defendant; therefore he should be sure to see who the violator is.

In those cases where the person refuses to carry out the requirements of the law and the interview with the municipal attorney has not produced the desired results, the inspector will ordinarily request that a complaint be issued. Sometimes the municipal attorney will have the complaint issued immediately following the hearing held in his office wherein the inspector and the violator were both present to tell their stories. In any case, it is the request for a complaint that is judged by the municipal attorney; he cannot issue one if the facts presented fail to establish a corpus dilecti (meaning that the specific crime charged has been committed by someone). Sometimes the request for a complaint may arise because of the arrest of the violator by the fire inspector. Whether or not it is customary to arrest violators of fire ordinances depends upon the policy of the particular department. Except in cases of clearly defined urgency, it is a somewhat risky procedure owing to the possibility of being sued for false arrest, although the likelihood of having a judgment rendered against him is rather remote. Moreover, in many cities the city attorney is only permitted by charter or the governing body to defend the chief and not the fire fighters of the fire department in such false arrest actions.

Authority to Make Arrests

With respect to the authority of fire fighters to make arrests, bring in witnesses, make necessary investigations and file complaints, several points should be noted. A member of the fire department has the same right that exists in any citizen to make an arrest. He may do so (1) whenever a public offense is committed or attempted in his presence; this power is sometimes exercised when, for example, a person is found to be smoking in a place that is forbidden by law, or is blocking the exits in a crowded place of public assembly; (2) likewise, when he finds that a felony, arson for example, has been committed and he has reasonable grounds for believing that the person arrested has committed the offense; or (3) when he knows that the person whom he is about to arrest has actually committed a felony, although not in his presence. In California the circumstances under which, and the time when, an arrest may be made by a peace officer and a private citizen are set forth in separate sections of the Penal Code. The two instances in which a private citizen is not accorded the power of a peace officer to make an arrest are: (1)—on a charge made, upon a reasonable cause, of the commission of a felony by the party arrested, and (2)—at night, when there is reasonable cause to believe that a felony has been committed by the person who is the subject of the arrest. Of course a peace officer can also make an arrest in obedience to a warrant delivered to him, but he cannot even do this at night for a misdemeanor unless the warrant so directs.[1]

How to Make an Arrest

An arrest is accomplished by placing the person under restraint or control or by his submission to the custody of an officer; he should not be subjected to any more restraint than is necessary for his detention. In making the arrest, the person must be informed of the intention to arrest him, of the cause of the arrest and of the authority to make it, except of course when the person to be arrested is engaged in the actual commission of the offense or is being pursued immediately after its commission. If the arrest is being made under a warrant, then it must be shown if required, and, if the person flees or forcibly resists, the officer may use all necessary means to effect the arrest, provided the person is a felon or the arresting officer has a warrant in the case of a misdemeanor. It is not proper to shoot a person who is fleeing from arrest, for example, if the offense committed is only a misdemeanor. If, in shooting above him to scare him into submission, the misdemeanant is accidentally injured, the person attempting such a method without a warrant could be held guilty of a criminal assault.

Informing a Suspect of His Rights

Although a discussion of the subject of the interrogation of witnesses has been previously presented in Chap. 5, a brief summary of the most important matters to be included in advising a suspect of his rights is reviewed here. Former Attorney General Thomas C. Lynch of California has recommended a method of informing criminal suspects of their constitutional rights. It has

---

[1] Provisions relating to power of private citizens to make an arrest can be found in West's Ann. Penal Code § 837.

gone to law enforcement officers in California, but is equally applicable to law enforcement officers throughout the United States.

His action is based on the U.S. Supreme Court decision, Miranda v. Arizona et al.[2] His statement: "We are faced with a situation in which the peace officer's initial approach to a suspect can basically affect the ultimate criminal conviction. Because of the Supreme Court's decision, we must be most careful and exact in adhering to constitutional requirements concerning pre-arrest, arrest, interrogation and investigation methods. The purpose of these recommendations is to ensure that criminals do not escape punishment because of constitutional legal errors made by arresting and interrogating officers."

The Specific Warning

"You have the right to remain silent.

"Anything you say can and will be used against you in a court of law.

"You have the right to talk to a lawyer and have him present with you while you are being questioned.

"If you cannot afford to hire a lawyer, one will be appointed to represent you before any questioning, if you wish one."

He also set forth specifically how to take the suspect's waiver of his right to remain silent and his right to counsel. The waiver is:

"Do you understand each of these rights I have explained to you?

"Having these rights in mind, do you wish to talk to us now?"

The Attorney General suggested that all officers carry these exact statements with them on cards.

Admonitions

In addition to the specific statements of rights and waivers, Lynch set forth nine admonitions based on the Supreme Court rulings, as follows:

1. All statements—with a few exceptions—are subject to warnings;

2. If warnings are not given, both the statements and evidence obtained through them are inadmissible;

3. A defendant must be warned regardless of his knowledge of his rights;

4. His waiver must be knowingly and intelligently made and his physical and mental condition must show that he is giving a "knowing and intelligent waiver";

5. Questioning must stop whenever a subject indicates his desire to remain silent;

6. Whenever a subject wishes to consult an attorney, questioning must stop;

7. If a suspect wishes to talk without counsel, his statement should be taped, if possible, or a hearing reporter should be present; if possible, the suspect should sign and date the statement;

8. If a waiver is signed, the peace officer should evaluate the suspect's education, mental and physical condition and any other factors to corroborate that it is a "knowing and intelligent waiver";

---

[2] Miranda v. Arizona, *et al*, 384 U.S. 436 (1966).

9. Lengthy interrogation or incommunicado incarceration before a statement is made is strong evidence of no valid waiver.

Exceptions to the Need for Warning

The ordinary traffic citation would not require a warning. No warning need be given in the following circumstances:

1. When a person walks into the police station and states that he wishes to confess to a crime;
2. when a person calls the police to offer a confession or any other statement he desires to make;
3. when the officer is engaged in "general on-the-scene questioning as to facts surrounding a crime or other general questioning of citizens in the fact-finding process. . . . It is an act of responsible citizenship for individuals to give whatever information they may have to aid law enforcement."

Interrogation—Search

If a fire investigator stops a man for questioning, a perfectly proper thing to do at the scene of a fire or similar circumstance, but the person refuses to answer any questions or explain what he is doing there, such refusal, in the absence of a statute to the contrary, is not considered a breach of the peace so as to justify placing him under arrest without a warrant. Nor can a person be arrested for one cause without a warrant, as, for example, for transporting fireworks illegally, and then when the search reveals no fireworks but the person is carrying a weapon, attempt to justify the arrest on the latter cause. Some jurisdictions do permit the use of evidence obtained in an unlawful search to justify the arrest where there was probable cause to believe that the search would result in finding incriminatory evidence. A California case so held, ruling that if the defendant wanted to object to the arrest he should have done so prior to the trial or upon a reasonable opportunity.[3]

The federal law does not permit, however, an exploratory search without a warrant for the purpose of getting evidence, and if a person is arrested, the right of search only ex ends to his person and does not justify taking him home and searching his house without a warrant.

Any person making an arrest may take from the person arrested all offensive weapons which he may have about his person, and must deliver them to the magistrate before whom he is taken.

Detention of Witness—False Arrest

A private person who has arrested another for the commission of a public offense must, without unnecessary delay, take the person arrested before a magistrate or deliver him to a peace officer. If the person is not placed under arrest, but voluntarily consents to go to an office for an interview this is proper

---

[3] Hill v. California, 401 U.S. 797, 91 S.Ct. 1106 (1971), but overruled by Chimel v. California, 395 U.S. 752, 81 S.Ct. 2034 (1969), because the search had been conducted before the Chimel decision. See Williams v. U.S., 273 F.2d 781 (9th Cir. 1959), cert. den. 362 U.S. 951, 80 S.Ct. 862 (1960); also U.S. v. Payne, 429 F.2d 169 (9th Cir. 1970, at 171).

as long as he is not told that he must submit to such questioning. There is nothing in the law that authorizes peace officers or anyone else to "bring in a witness" with legal process for the purpose of securing testimony or information. The forceful detention of a person not lawfully arrested constitutes a false arrest for which peace officers as well as private citizens may be held liable in a civil suit. Moreover, the failure to take the arrested person promptly before the proper official is usually declared by law to be a misdemeanor.

Warrants

A warrant of arrest is an order of the court issued by the judge. If the warrant is valid on its face, the officer is protected from a false arrest suit provided he has made the arrest properly. Incidentally, a "John Doe" warrant for a man's arrest must have attached a description of him where his name is not known; it is not proper to arrest a man on a warrant that was not specifically issued for him.

Malicious Prosecution

The filing of a complaint against a person, or commencing an action against him, whether civil or criminal, without probable cause (reasonable grounds) is considered a "malicious prosecution," for which a civil action can be brought for damages.

Arson

Some fire departments are charged with the responsibility for making investigations of suspicious fires, interviewing witnesses, collecting evidence, arresting suspects, and assisting the prosecutor in the preparation of cases for trial. Since there are several books which have been published on this subject, the discussion here will be limited to a few basic observations of a legal nature. Relevant topics, such as search and seizure, interrogation of witnesses, waivers and warnings, arrest procedure, etc., have already been discussed in this chapter.

At common law, arson was the felony of maliciously and wilfully burning the dwelling house of another person. It was an offense against the security of a man's habitation, and therefore was considered a crime of greater enormity than other unlawful burning. It was an aggravated felony because it showed a greater recklessness and contempt of human life than the burning of any other building in which no human being was presumed to be living. The crime is now statutory in most jurisdictions, and though some penal codes define the crime substantially the same as it was at common law, some statutes make the crime applicable to barns, hayfields, bridges, and all kinds of property.

Elements of the Offense

The two elements which must be proved beyond a reasonable doubt in order to sustain the conviction for the crime of arson are: (1) The *corpus delicti;* that is, a fire caused by a criminal agency; and (2) the identity of the defendant as the one responsible for the fire. The mere burning of a structure does not establish the *corpus dilecti,* for there is a rebuttable presumption that a fire is of accidental origin. The burning, to constitute arson, must be the result of the wilful and criminal act of some person; even though there may be prima facie evidence that a fire was of incendiary origin, such as the finding of separate and distinct fires on the premises,[4] and even though only one person was known to have been on the premises at the time of the fires, there is a presumption of innocence accorded to that person which would have to be rebutted in order to provide the second element of the crime of arson.

Circumstantial Evidence

The very nature of the crime of arson, which is one of secret preparation and commission and seldom observed by eyewitnesses, renders it necessary for the state to establish the guilt of the accused, in many, if not in most, cases, by relying on circumstantial evidence. In a Wisconsin case,[5] a verdict of guilty had been found by the jury against a man who had been seen leaving the house of a girl friend just before the fire was discovered and shortly after he had, in a fit of anger, announced that he was going to burn her house to the ground. In sustaining the verdict, the court stated that circumstantial evidence properly established may be and often is stronger and more convincing than testimonial evidence. While proof beyond a reasonable doubt must measure up to the same standards whether the evidence is direct or circumstantial, the same established tests must be applied on appeal to measure the sufficiency of the circumstantial evidence, i.e., whether the evidence adduced, believed, and rationally considered by the jury was sufficient to prove the defendant's guilt beyond a reasonable doubt. In referring to the defendant's argument that the fire was not proved to be accidental, the court said, "While as a matter of conjecture the remote possibility of a fire in the farm dwelling, occasioned by natural causes, was not ruled out, the evidence did so; hence that speculative concept could not avail the defendant, the law being well-settled that a jury must act on probabilities, not possibilities, and the mere possibility that a fire was occasioned by spontaneous combustion or by some other cause innocent of criminal intent, did not demand an acquittal."[6]

Intent and Motive

In a legal sense, intent is quite different from motive, and may be defined as the purpose to use a particular means to effect a certain result, while motive is the reason which leads the mind to desire that result.[7] Intent is ordinarily an inference of law from acts proved (Ballentine, *Law Dictionary*). Ballentine's also defines "criminal intent" as "that evil state of the actor's mind, accompanying an unlawful act, in the absence of which no crime is committed."[8] As

---

[4] People v. Sherman, 97 Cal. App.2d 245, 217 P.2d 715 (4th Dist. 1950).
[5] State v. Kitowski, 44 Wis.2d 259, 170 N.W.2d 703 (1969).
[6] Grimes v. State, 79 Ga. App. 489, 54 S.E.2d 302 (1949).
[7] Baker v. State, 120 Wis. 135, 97 N.W.2d 566 (1903).
[8] Ballentine, James A., *Law Dictionary with Pronunciations,* Lawyers Co-operative Pub. Co., Rochester, New York, 1930.

pointed out in Chap. 1, in the discussion of criminal prosecutions, in every crime there must exist a union, or joint operation, of act and intent, or criminal negligence. In a prosecution for arson, the criminal intent may be inferred from the defendant's acts, statements, or circumstances connected with the offense.

Motive, in criminal law, according to Ballentine, is "that which leads or tempts the mind to indulge in a criminal act."[9] It is an inferential fact, and may be inferred not merely from the attendant circumstances, but in conjunction with these, from all previous occurrences which have reference to, or are connected with, the commission of the offense. Motive, then, is the cause or reason which induces a person to commit a crime, and while its determination may be helpful in establishing the guilt of the defendant, it is never legally necessary to prove motive in order to obtain a conviction. Arson investigators must look for motive, however, for it is useful to help explain the actions of the parties involved, and even to connect the defendant with the offense charged.

What is "Burning"?

To have the crime of arson there must be a "burning" of something. The crime is complete in this respect if the nature of the material fired is changed or charred, according to Ballentine, who cites 2 Ruling Case Law 496.[10] A wasting of the fibers of the wood, no matter how small in extent, is sufficient to prove arson if the other elements of the crime can be established.[11] "The law is well established that a charring of the fibers of a part of a building is all that is required to constitute a burning sufficient to make the crime of arson. Any charring is sufficient." When the question was raised whether the acoustical tile is part of the building, the court said, "We cannot see any reason to make a distinction between a wood panel ceiling and an acoustical tile ceiling. Each is an integral part of the building."[12]

Polygraph Evidence

The care which should be taken in obtaining confessions so as to make them admissible into evidence has already been discussed in this chapter. On the question of the admissibility of polygraph tests in criminal trials, the courts have generally ruled against them, a practice followed by the U.S. District Judge in San Diego, Gordon Thompson, Jr., in 1972, in the case of U.S.A. v. Bruce Eugene de Betham.[13] After listening to four days of testimony by national experts on polygraphy, including the director of the U.S. Army's Criminal Records Division, the judge asserted that he had been impressed by evidence that polygraphy had indeed gained in reliability over the past years, and said that under certain controlled circumstances the validity of the tests might be assured. He also said that he was impressed by Army expert testi-

---

[9] *Ibid.*
[10] *Ibid.*
[11] People v. Oliff, 361 Ill. 237, 197 N.E. 777 (1935).
[12] State v. Nielson, 24 Utah2d 11, 474 P.2d 725 (Utah Sept. 28, 1970).
[13] United States v. De Betham, 348 F. Supp. 1377 (SD Cal. 1972), cert. denied, 412 U.S. 907, 93 S. Ct. 2299 (1973).

mony that in only two cases have suspects who have passed the polygraph been subjected to criminal prosecution. These two were the colonel and captain who were tried in relation to the My Lai massacre in Vietnam. However, the judge said he was bound by two 1971 Ninth Circuit of Appeal cases to refuse the tests to be admitted into evidence. These cases were U.S. v. Sadzadeh, 440 Fed.2d 389, and U.S. v. Salazar Gaeta, 447 Fed.2d 468, neither of which developed facts or detailed reasons for the exclusion of such evidence, and Judge Thompson refused to speculate on what might have been the reasoning of the circuit court.

At the present time, polygraph tests are generally admitted only upon the stipulation of both parties. They are frequently used in California state trials upon stipulation, and have been especially instrumental in the disposition phase of a criminal case. A California Superior Court judge who uses the tests frequently in child molestation matters in the question of granting of a motion for new trial, said that the reliability of polygraphs is almost totally dependant upon the character and experience of the operator. It is a science, he said, but one much like psychology, in which the "expertness" of the expert is sometimes open to question.[14]

Handling "Sit-ins"

Because policemen are sometimes reluctant to clear the exits of restaurants and similar places of public assembly, for fear of being accused of "police brutality," fire fighters are sometimes called upon to handle these situations. It is a good practice for the fire prevention officer to respond with a policeman and a photographer, and if the warning of possible arrest results in clearing the aisles, then no other action need be taken. If this is ineffective, then a head count should be made of everyone in the restaurant, for example, which is overcrowded and in which the sit-in is being staged.

The fire officer should then announce two or three times, using a voice amplifier, that the offenders are breaking the law, giving the ordinance number. He should warn them that they will be arrested if they do not move. If they do not, he should announce that they are all under arrest and request the police to carry them off to jail. The police can use reasonable force in restraining the persons arrested, and if they resist arrest, this resistance action becomes a felony. Remember that the exits blocked must be *required* exits, and be sure to obtain the names of witnesses who will testify as to the announcements made. The photographs will be useful in court to establish the violation.

Carrying a Police Badge

The mere fact that a fire fighter carrys a police badge does not make him a peace officer within the meaning of the penal code definition, and the failure to

---

[14] Adler, Marlene, "Polygraph Evidence Not Admissible," *League News* (Los Angeles Police and Fire Protective League), Aug. 1972, p. 3.

carry a badge has relatively little effect upon his powers or activities. Nor does an ordinance giving him the powers of a "police" officer necessarily bring him within the state law's definition of a "peace" officer, where such law specifically sets forth certain classes of enforcement personnel and omits any reference to fire fighters.

Signing the Complaint

A criminal complaint is signed by the person making the complaint in his capacity as an individual and not as an official. Legally it is immaterial whether a complaint is sworn to by a police officer or a private citizen. Therefore the members of the fire department stand on the same footing as do members of the police department with respect to formally filing a criminal charge.

Hearings before the City Attorney

In hearings before the city attorney it is best for the fire inspector to be impersonal; he should just state the facts and attempt to avoid giving the violator the impression that his prosecution is a personal issue with him. He should not argue with the citizen who has been ordered to appear at the hearing.

Trial Preparation

It is important that his statement of facts be as complete as possible, and should be based upon the same information which he has submitted to the head of the fire prevention bureau or fire chief for submission to the city attorney in the request for the issuance of a complaint. Most prosecutors never see the file on a case until it comes to trial—usually his first look is while the judge is taking the bench and the first witness is being called. Therefore all the facts must be stated. Conversations with the violator should be included if these constitute admissions; as such they would be admissible in evidence. If the violator cursed the inspector, the exact words should be used in the statement, so that if he pleads guilty and the judge looks at the complaint to determine the fine, the man will get his just desserts. However, before repeating profanity or other testimony which may be embarrassing to jurors, alert them to this fact before you testify to it.

Sometimes it is difficult to ascertain the name of the party to whom a notice should be directed. Where the business has been issued a license or permit by any agency, use the name listed on any such permit found posted on the premises in question; if the name listed is "Jane Doe," and a "Jane Doe" is in the court room, then the presumption is that they are one and the same person and the burden shifts to the defendant to prove otherwise.

Past acts which might indicate motive, design or intent can be shown.

Evidence of previous convictions of felonies can be used for the purpose of impeaching the credibility of the testimony given, and past offenses may also be admissible to show design or intent, or an overall plan or scheme.

An inspection should be made the day before the arraignment because if the person pleads guilty to an offense which took place say ninety days before that time, the judge will want to know the present status in determining the fine; if the situation has already been remedied the fine is likely to be light.

In the Courtroom

The inspector who has been working on the case should act as an advisor to the prosecutor in the courtroom; he can assist him not only on the fire ordinances themselves, but also on technical terms, such as "flash point," "spontaneous ignition," "exposure," etc.

On the Witness Stand

More often than not the case will be tried before a jury. The inspector's appearance is important; he should wear a freshly pressed uniform. The witness should also be very impartial, and particularly careful not to give the impression that he dislikes the defendant; it is better for him to say, "I don't know the defendant, but I am familiar with the facts, etc." Then he should testify only as to the facts; he should not give his opinions unless he is called in a case as an expert witness, in which situation he will be given a hypothetical question and a proper foundation will be laid for the introduction of his opinions, e.g. "How many years have you spent in fire investigation, or fire inspection? Have you had an opportunity to observe many situations of this nature?" etc. (to show his qualifications to give an opinion).

"Horse-Shedding" the Witness

The above expression dates back to the days when the lawyer took his client out to the horse shed just before the trial and went over with him the questions he was going to ask him on the stand; this would give a last minute chance to see if he had learned the right answers.

It is perfectly proper for the fire department to go over the testimony that an outside witness is requested to give in a case. This does not mean that he should be told to color his story or change his viewpoint. It means that he should be coached to the point where he will not give purely heresay evidence, nor tell of things that are immaterial or irrelevant to the case.

He should be told that if the lawyer for the defendant asks him whether or not he "discussed this case with anyone," to answer, "yes"; if he blurts out, "Oh no, Sir," thinking that was the right answer, and then if the same lawyer asks him whether or not he discussed it with the fire officials or city attorney, he would have to give the contradictory answer, "yes," and thereby possibly become rattled.

Hearsay Evidence

The fire department witness should never volunteer information; it is a very dangerous thing to do, for something might be inadvertently referred to which would prejudice the case. Nor should he make speeches. However, if a question calls for a "yes" or "no" answer he has the right to request the court's permission to explain his answer. He should avoid giving any *hearsay* evidence, i.e., information which he obtained from someone else; only facts derived from his own knowledge or observation are generally admissible in evidence for the reason that the defendant has the right to cross examine anyone who says something about his case; if the inspector quotes someone who is not there the defendant has no chance to question that person directly about the information; if that person is present in the court then he should be called to testify to such facts if they are material to the issue.

The "Know-It-All"

Nor should the witness attempt to give the impression that he knows all the answers. The jury is likely to distrust a person who is too glib and too ready with a positive answer to difficult or technical questions. If the attorney for the defendant should ask the fire department investigator, for example, "What is the temperature at which nitrocellulose commences exothermic decomposition?" or "How many minutes will a 'king size' cigarette burn on top of a newspaper before being completely burned to ash?", if he doesn't know the answer he should frankly say so, or say he doesn't remember; and unless he is called as an expert witness (and as such, should be prepared) he should not even venture a guess as to the answer to such a question.

Avoid Specialized Expressions

Avoid the use of specialized vocabulary or "canned" expressions that sound like a reiteration of lecture notes from the training academy. For example, a police officer hardly ever gets out of his cruiser, he usually "alights" from his "vehicle" and rarely walks to anything; he generally "proceeds." Plain English is best.

Silence Is Not Always "Golden"

There are times also when the witness's silence may have its implications—wrong implications; the judge or jury may draw the wrong inference from such silence. For example, at a coronor's inquest following a hotel fire, the fire chief may be asked the date of the last fire prevention inspection; to give no answer may result in more than an adverse reflection upon the fire department's records.

Exhibits and Notes

It is good practice to bring in physical evidence, exhibits, etc., to use a blackboard, draw diagrams, and by dramatic illustration demonstrate physical acts which tend to prove the case. Photographs introduced with the proper foundation are also excellent aids. Any notes which may have been taken should also be brought to the court room, for these may be used to refresh the memory with permission of the court; the witness does not read his notes out loud, but after refreshing his recollection, he can testify without them. They can also be admitted in evidence as "past recollection recorded," where, even though the events occurred so long ago that they do not refresh his memory, yet they were made by him at the time when the facts were fresh and he can testify as to their accuracy. However, a witness is not required to produce notes used to refresh his memory unless he actually looked at them while testifying.

Sequestration of Witnesses

It should be remembered that the court can sequester the witnesses, i.e., require all but the one who is to testify to leave the room, so as to prevent subsequent witnesses from being able to pattern their testimony to fit in with what the prior one has stated; if the court sends the other inspector out of the room the one on the witness stand should confine his testimony to the facts, and not say the wrong thing—a good piece of advice in any case.

Impeaching a Witness

Another reason why the keeping of accurate notes is important is to be able to take care of a situation where a witness for the prosecution gets on the stand and tells a different story from what the inspector thought he would; in such cases the prosecution can impeach his own witness if he gives surprise testimony which contradicts another witness who had taken careful notes. In this connection it is well to reiterate that if the inspector uses notes, such notes may be required to be shown to the jury by the counsel for the defendant, and if there is any inconsistency between what the inspector has testified to and what he has in his notes, the effect upon the jury may be detrimental to the case.

Corpus Delicti

This latin phrase, which literally means, "body of the crime," when used in the legal sense, means that the specific crime has been committed by someone. It is made up of two elements: first, that a certain result has been produced, as that a man has died or that a building has been burned; second, that someone is criminally responsible for the result. In a typical case, it might be established by asserting that "a man was smoking; he was in a film vault" (where this is prohibited by law). The allegation that a man was smoking in a film vault establishes a corpus delicti, even if this act has not yet been tied in with the particular defendant.

Relevancy

Do not put into evidence anything which is not relevant to the case. If the violator says to the inspector, "Do not file on me because I'm on parole, and will go back to the pen," do not include this in the testimony, for if it is brought out on the stand the court is likely to declare a mistrial for the introduction of prejudicial matter. Therefore, while it is proper to testify that a conversation was had with the defendant, confine the testimony as to what he said to matters which are strictly relevant, i.e., have a bearing on the issues involved; if, on cross-examination, the question is asked by the defendant's attorney whether or not any other conversations were had with the defendant, then it is proper to say, "yes" (if such was the case), and if he asks what they were, it is all right to relate them.

Do Not Stress the Immaterial

In addition to avoiding testimony as to facts which are not relevant to the case, any undue emphasis should not be put upon matters which are not really material to it. For example, it may be relevant that a defendant was storing high-octane gasoline in his warehouse, but when the charge is unlawful storage of flammable liquids, it is not really material whether the liquid was gasoline or benzene, much less whether it was "high" or "low" octane and any detailed description of the type of flammable liquid, brand name, etc. may detract from the more important information you wish to present. The main point you may desire to make is that the liquid had a flash point which brought it within the code definition of "flammable," a fact that could have easily been established in a laboratory, and thus obviate the necessity of proving whether the liquid was a specific chemical or product.

"Sticks and Stones"

The witness should never be argumentative in his testimony, nor should he allow himself to get mad at the attorney for the defendant; that is what he would like. It is best to take plenty of time and answer each question deliberately; the effect is more likely to needle the attorney than the witness.

The Verdict

With respect to the decision handed down in the case, it is not wise to "take it to heart" if the case is lost; it is not fitting for a public official to act like a poor loser; the fact that the jury decides the issue against the city does not necessarily mean that it thought the inspector or other witnesses for the prosecution were liars. Consider the case a worthwhile experience if even one good lesson was learned. Analyze the case to see where it might be improved if a similar one were to be presented again; perhaps a little more attention to the basic field work and a little better preparation might be the winning factors next time.

Epilogue—"But Is It Fair?"

In attempting to arrive at a course of action, for example, whether to reprimand or suspend a member for violating a rule; whether to prosecute a businessman for a code violation or to grant him a requested extension of time to comply; whether to grant or deny an injured member a leave or pension, do not merely consider the legal aspects of your decision, but follow the example of the former Chief Justice, Earl Warren, who often interrupted legal arguments with the question: "But is it fair?"

That question is basic to any system whose end is justice, and notwithstanding all the cynicism which has arisen regarding the government and the law following the Watergate hearings, the "name of the game" is still "justice"! Although a strict adherence to the law may often bring about a just result, in the sense that the parties involved are accorded the treatment which the law permits or requires, such results may nevertheless appear to be highly arbitrary and unreasonable to a fair-minded person. Most people want more than justice—they want to be treated fairly!

For example, your superior officers may use their authority to change your assignment from the engine house to the fire prevention bureau without consulting you, or they may transfer you to the "big house," when your seniority would normally let you relax in a residential station, without even the gesture of asking your preference, but is it fair? The city council or board of supervisors may exercise its power to withhold a much needed pay raise to combat the ever-rising costs of government, but is it fair? A fire fighter's union might demand a thirty-six hour work week and a 100 percent pension after thirty-five years of service, while other employees in the same city may have to work a forty-hour week for forty years to obtain a fifty percent pension; through the exercise of political pressure the union might secure its demands, but is it fair?

Because of the proliferation of damage claims, malpractice suits, and the ever-increasing consciousness of the possibility of being faced with litigation, there is a growing caution about offering assistance and voluntary acts of kindness in emergency situations on the part of persons who are qualified to render aid to the unfortunate. There is no law which compels a doctor, paramedic, or first-aider to stop while off duty and render assistance to accident victims. They can all legally drive right on past the scene without stopping to lend a hand, but is it fair? They have all been especially trained, often at public expense, to be able to assist others in time of peril, and no doubt their inner impulse would be to offer help. But too often they curb this kind inclination because they are afraid of some untoward result. Never let the fear of an ingrate's lawsuit deter you from the humanitarian service that you are capable of offering. In the unlikely event that you should ever have to face a judge or jury of your fellow citizens, the most important question which they will have to ask themselves about the decision in your case will be, "Is it fair?" If *you* have been fair in your treatment of *others,* why not take your chances on their treating you fairly when the chips are down?

A hundred books could be written about legal principles, but the greatest principle that was ever propounded, by the greatest man who ever lived, is the one you should always follow when in doubt, and it simply requires that you do unto others what you would have them do unto you. No one who follows that admonition need ever fear the judgment of court or jury, for that rule is exceedingly fair!

Charles W. Bahme

Review Activities

1. (a) What two elements in an arson-related crime must be proved beyond a reasonable doubt?
   (b) Does your community have any programs to combat the problem of arson? If so, describe one such program. If not, research the arson-control program of any large city and describe it.
2. Differentiate between the meanings of "intent" and "motive."
3. In your own words, write a definition of the term "corpus delicti."
4. When does a fire fighter have the authority to arrest someone? How is the arrest made?
5. Describe the circumstances during which a fire fighter has the authority to break up a "sit-in."
6. (a) Why is the fire inspector's manner and appearance important?
   (b) What is the fire inspector's best method of recommending safety changes to an owner of a building?
   (c) Outline the steps taken if a building owner does not comply with a fire inspector's "Recommendation Memorandum."
   (d) Discuss the fire inspector's conduct when appearing as a witness in the courtroom. Then compile a list of "do's and don'ts" for fire inspectors when appearing on the witness stand.

GLOSSARY OF LEGAL TERMS*

ab initio. "From the beginning."

abandonment. As applied to property, it means the voluntary relinquishment of the possession of a thing by the owner with the intention of terminating ownership.

acknowledgment. An admission or confirmation; another meaning is the public declaration that the act evidenced by the instrument is his act or deed.

absolute. Unconditional; unrestricted.

absolute liability. Liability without fault; discussed on p. 125.

accessory. See p. 13.

accessory after the fact. See p. 13.

accessory at the fact. See p. 13.

accessory before the fact. See p. 13.

accord and satisfaction. An agreement between two parties to give and accept something in satisfaction of a right of action which one has against the other, which when performed is a bar to all actions upon this account.

action. An ordinary proceeding in a court of justice; the right to sue or bring an action; a judicial remedy.

actionable. Remediable by an action at law.

admissible evidence. Evidence which a court or tribunal may properly receive for consideration in the deciding of a case.

admission. A statement made in conflict or against the proprietary or pecuniary interest of the person making it.

affidavit. Any voluntary written statement of a person sworn to or affirmed before a notary public or other officer designated by law to administer an oath or affirmation.

agent. A representative vested with authority to create obligations for his principal.

alias. Otherwise; "also known as."

allege. To plead or make an assertion.

alter ego. The "Other I," or counterpart.

amicus curiae. "A friend of the court"; includes both laymen and lawyers who are requested to advise the court in a judicial proceeding before it.

---

*Where a term has been defined completely in the text, the page reference is given.

GLOSSARY OF LEGAL TERMS 245

animus. Mind; intent.
answer. A plea interposed by the defendant to a declaration or complaint; the defendant's written response to the plantiff's complaint.
ante. Before.
ante-date. To insert a date that is prior to the actual date of the instrument's execution.
appeal. Any complaint to a superior court of an injustice done by an inferior one.
appellant. A person who files an appeal. See p. 6.
appellate jurisdiction. Jurisdiction to retry a case that has already been tried in some other tribunal. See p. 16.
arbitrary exercise of privileges. See pp. 80–81.
arraignment. See p. 21.
arrest. The detention of a person or taking him into custody; no actual touching of his person is necessary, for it is sufficient if the person being arrested understands that he is in the power of the arresting party, and submits thereto. See pp. 152, 229, 230.
arson. At common-law it was the wilful and malicious burning of the dwelling house of another; its meaning has been broadened by statute in many jurisdictions to include one's own house and the appurtenances thereto.
assignment. The transfer of a property right from one person to another. See p. 181.
assumption of risk. Usually applied to the hazards of an employment which are so obvious to the servant that the master is under no duty to warn him of them. See pp. 97, 124.
attorney in fact. A private person who has been authorized by his principal to do a particular thing or otherwise act in his stead. A private attorney, as distinguished from an attorney at law.
attractive nuisance doctrine. The view that one who maintains contrivances on his premises of a character likely to attract children in play, or permits dangerous conditions to remain thereon knowing that children are in the habit of playing around them, is liable to a child injured therefrom. See p. 99.
bail. See p. 21.
bona fide. Good faith; without fraud or collusion.
bribery. See p. 7.
brief. A detailed statement of a party's case.
business visitor. See pp. 124, 201.
case. A contested question in a court of justice.
cause of action. See p. 8.
caveat emptor. Latin—"Let the buyer beware."
certiorari. A common law writ used to get a court to review a decision of a lower court. The Supreme Court of Oklahoma ruled in 1944 in a case involving the removal of a police officer that this writ could be used for the purpose of bringing up only two questions: (1) Did the inferior court or board have jurisdiction, and (2) Did it keep within that jurisdiction in

the order made or action taken; that it cannot be used to correct errors of law or fact committed by the lower court or board within the limits of its jurisdiction. See pp. 19, 196.

circumstantial evidence. See "evidence."

cite. See p. 6 (footnote).

civil action. Civil suit; an action brought to enforce a civil right, as distinguished from a criminal prosecution. See p. 6 (footnote).

code. See p. 2.

common law. See p. 2.

complainant. See p. 7.

complaint. Civil, see p. 20; criminal, see pp. 21, 237.

compounding a crime. See p. 7.

condemnation. The exercise of the power of eminent domain (q.v.).

condition precedent. A condition which must happen before the right will accrue.

condition subsequent. A condition, by the failure or nonperformance of which, a right already vested may be defeated.

confession. The voluntary admission or declaration by a person who has committed a crime or misdemeanor, to another, of the extent of his participation in it. See p. 154.

confiscate. To appropriate property; to transfer property from a private to a public use, e.g., property of an enemy or contraband.

constitutional law. See p. 1.

constitutional right. See p. 1.

constructive notice. Notice imputed to a person not having actual notice. See p. 123.

contra. Otherwise; disagreeing.

contract. An agreement by which a person undertakes to do or not to do a particular thing.

contributory negligence. Such negligence on the part of the plaintiff as to make the injury the result of the united, mutual, concurring, and contemporaneous negligence of the parties. See pp. 124, 221.

corpus delicti. The body of the offense; the essence of the crime, see p. 233.

corporate function. Meaning "private," as distinguished from the word "public"; sometimes used in connection with the acts of a governmental subdivision to refer to those which are proprietary and nongovernmental in nature. See p. 63.

crime. An act committed or omitted in violation of public law either forbidding or commanding it. See p. 11.

criminal law. See p. 10.

criminal negligence. See p. 12; see also "negligence."

criminal offense. See p. 12.

damnum absque inuria. "A wrong without a remedy"; see pp. 8, 116.

decision. A judgment given by a competent tribunal; should not be confused with "opinion," which represents the reasons given by the court for its decision.

defendant. See p. 6.

deposition. A written declaration under oath, made upon notice to the adverse party, for the purpose of enabling him to attend to cross-examine.

demurrer. A pleading which admits all matters of fact which are sufficiently pleaded; a general demurrer takes exception to the sufficiency of the previous pleading in general terms without showing specifically the nature of the objection. See p. 20.

detention of witness. See p. 232.

dictum. (Also "Obiter dicta.") A view expressed by the court, but which, not being necessarily involved in the case, lacks the force of an adjudication.

direct evidence. See "evidence."

directed verdict. A verdict which the jury returns as directed by the court; the court takes over when there is no competent, relevant, and material evidence to support the case, or where the evidence is contrary to all reasonable probabilities, or where it is clear that any other verdict would not be allowed to stand; in such cases, the court tells the jury what it must bring in as a verdict.

doctrine of . . . . See the other caption, e.g. "assumption of risk," "fellow servant," "respondeat superior," "stare decisis," etc.

due process of law. Law according to the established safeguards for the protection of private rights. See p. 119.

easement. The right which one person has to use the land of another for a specific purpose.

emergency call. See p. 78.

eminent domain. The superior right of the government to take private property in certain cases or control its use for the benefit of the public by the payment of the reasonable value thereof, as distinguished from its right to take private property under its police power or by taxation.

enjoin. To command or require a person to do or not to do a particular thing; in equity this is done by a writ of injunction.

entrapment. See p. 152.

equity. In its broadest meaning it signifies natural justice, but also refers to a branch of remedial justice by and through which relief is afforded to suitors in the courts of equity; distinctions between actions in law or equity have been abolished in many states today, but the remedies, though sometimes having a different name, are usually still available. One of the old maxims was, "He who seeks equity must do equity."

escrow. A deed delivered to a stranger, to be delivered by him to the grantee upon the happening of certain conditions, upon which last delivery the conveyance of the title is complete.

estoppel. The doctrine of estoppel (or "estoppel in pais") rests upon the principle that where a person wrongfully or negligently by his acts or representations causes another who has a right to rely upon them to change his position or condition to his detriment, said person is precluded from pleading the falsity of his acts or representations for his own advantage or from asserting a right which he otherwise might have had. The general rule is that the above doctrine will not be applied against the state and its political subdivisions when not acting in a proprietary capacity, unless clearly necessary to prevent manifest injustice. For example, the Supreme Court of Washington held in 1945 that the City of Bremerton was not estopped from pleading that it was not liable to a person who was injured by the negligent operation of a city truck on the basis that the collection of garbage was a governmental function. "Appellant was severely injured, apparently through no fault of his own; the trial court correctly ruled that under the circumstances of the case the law affords him no remedy." See p. 98.

evidence. That which tends to prove or disprove any matter in question, or to influence the belief respecting it. *Direct* evidence is that which applies immediately to the fact being proved, as for example if A saw B set his house on fire. *Circumstantial evidence* is that which usually attends other facts sought to be proved, e.g., where A found that his house had been set on fire by oil-soaked rags which he proved came from B's premises and B, an old enemy of A, had been seen running away from A's house just before the fire broke out; in the latter case, the jury might infer (or find the *inference*) that B set A's house on fire from the *indirect* or circumstantial evidence. See pp. 125, 126.

executed contract. One which has been fully performed; the Statute of Frauds (requiring certain contracts to be in writing) does not apply to an executed contract. See p. 179.

ex contractu. Arising out of a contract.

ex delicto. Arising out of a tort.

execution. In civil actions, it is the mode of obtaining the debt or damages or other things recovered by the judgment. See p. 20.

ex parte. On or from one side only.

ex post facto law. A law which makes an act punishable in a manner in which it was not punishable when it was committed, or which makes an act a crime which was not criminal at the time it was performed. See pp. 11, 119.

false arrest. An unlawful physical restraint by one person on the liberty of another, whether in prison or elsewhere. It does not require the actual touching of the person arrested, but merely that he be made to understand that he is within the power of the person making the arrest. See "arrest" and p. 232.

felony. A crime so declared by statute, or considered as such at common law. See p. 11.

fellow servant rule. The common law doctrine that where the employer, or master, provides a safe place to work, he is not liable for an injury to one of his servants caused by another servant's use or misuse of any of the tools or appliances. "Fellow servants" are all who serve a common master and who take the risk of each other's negligence. See p. 97.

governmental function. An undertaking by a political subdivision which is essential to its existence as such, e.g., maintenance of public buildings, fire and police departments, as distinguished from private functions, e.g., water and power system, street car system, etc. The distinction is important because the general rule is that a municipality is not liable for injuries in a civil action arising out of the performance of a governmental function. See p. 96.

habeas corpus. Latin for "you may have the body"; a writ directed to the person detaining another and commanding him to produce the body of the prisoner at a certain time and place and show cause why he should not be released. It is often called the great writ of liberty. See p. 17.

hearing. See pp. 192, 237.

hearsay evidence. Statements coming from a person who is not a party to the proceeding and not made under oath; such evidence is not admissible in a court of justice. See p. 239.

home rule. See p. 42.

impeach. To discredit the credibility.

incriminate. See p. 10.

indictment. See p. 21.

inference. A permissible deduction from the evidence. See p. 126.

information. See p. 21.

injunction. A writ forbidding a person from doing a noncriminal act, but one which is inequitable or against conscience.

inherent improbability. Refers to testimony which is so full of inconsistencies and contradictions that the court is warranted in rejecting it notwithstanding its acceptance by the jury.

interrogation. See p. 232.

in personam. See p. 16.

in re. "In the matter of."

in rem. Against a thing, and not against a person; about a thing. See p. 16.

inter se. Among or between themselves.

intra vires. Within the powers; within the scope of the authority granted by law. See pp. 109, 215.

invitee. One who comes upon the premises by the express or implied invitation of the owner. See pp. 124, 202.

judgment. The final determination of a court of competent jurisdiction upon matters submitted to it in an action or proceeding.

judicial council. See p. 16.

judicial notice. Is said to be taken of facts which are deemed by their very nature to be already known to the court and jury and which therefore do not have to be proved; e.g., that July fourth, 1952, fell on a Friday. It will not however take j.n. that a paint store under a hotel is inherently hazardous.

judicial system. See p. 16.

jurisdiction. The right to adjudicate concerning the subject matter in a given case. See p. 16.

justice. See p. 6.

laches. The doctrine of equity that a man cannot "sleep upon his rights" for an unreasonable length of time and then come into court and ask that he be given a remedy; it is grounded upon the principles of another equitable doctrine, "estoppel" (q.v.); the rule requires that a man who has grounds for an action, e.g., where he has been wrongfully removed from his position, be prompt in asserting his cause. Thus, undue delay in presenting a claim for a pension, or for reinstatement to a position, is said to be barred by laches. The Supreme Court of Washington in 1944 held that a City of Aberdeen policeman's mandamus action to compel reinstatement was not barred by laches in any period short of the statute of limitations unless it is made to appear that, by reason of delay in asserting the claim, the other party has altered his position or has otherwise been injured by the delay. See p. 166.

latent defect. A hidden defect; one which is not obvious; not patent.

law. See p. 1.

lawsuit. An action or proceeding in a civil court. See p. 6.

liability. Citizen's, see p. 121; city's—to fire fighters, see p. 106; and to the public, see p. 63; fire inspector's, see p. 142; fire fighter's, see p. 212; public's—to fire fighters, see p. 201.

libel. See p. 6.

licensee. A person who is licensed; a person who is on the premises of another, not by invitation, but by permission, or whose presence is merely tolerated. See pp. 124, 201.

line of duty. See p. 163.

malfeasance. See p. 8.

mala in se. With reference to offenses, it means those which are bad in themselves, as distinguished from those which are merely mala prohibita.

mala prohibita. Acts which are forbidden by statute but which are not otherwise wrong or wicked.

malice. Evil intent; an act done with malice is done intentionally and without justification. It does not necessarily imply an ill will toward a given person, but denotes that condition of mind which is manifested by the intentional doing of a wrongful act without just cause or excuse.

malicious prosecution. The commencing of any action or proceeding against another without probable cause and with malice. See p. 233.

mandamus. The remedial writ used to correct those acts and decisions of administrative agencies which are in violation of the law, where no other adequate remedy is provided. This writ is generally used to compel performance of ministerial duties, and in a few states has been used to review the final acts of boards and administrative agencies which do not exercise judicial power.

For example, it was used to review the action of the Los Angeles Civil Service Commission to test the proper exercise of discretion of that Board in sustaining an employee's discharge, i.e., to see whether the Board acted arbitrarily and whether there was substantial evidence to support its determination of fact. The general rule, however, is that mandamus is used to compel action and not to review it. (Certiorari is the writ ordinarily used to review an action.) See pp. 17, 198.

master-servant relationship. The relation which exists between two persons when one of them may control the work of the other and direct the manner in which it shall be done.

ministerial duty. A duty which is so definitely prescribed by law as to leave no discretion in the officer upon whom the duty is imposed. See p. 218.

ministerial function. Those functions of a municipal corporation which are the private corporate functions as distinguished from the public "governmental functions" (q.v.). See pp. 64, 70.

misdemeanor. Any crime which is neither punishable by death nor by imprisonment in a state prison; a crime that is not a felony. The courts have declared that where a crime may be punished either as a misdemeanor or a felony, it will be considered to be a misdemeanor only. See p. 11.

misfeasance. The improper doing of an act which one might lawfully do. See p. 8.

municipal corporation. Territorial and political subdivisions established by the state for the purpose of administering local government; a chartered city.

negligence. The failure to act as an ordinary prudent and reasonable man would act under like circumstances. See pp. 202, 221.

negligence per se. Negligence in itself; or negligence as a matter of law. See p. 128. Under the common law, the neglect of a legal duty imposed by statute or the common law constituted criminal negligence; however, in states which have abolished the common law, the bare neglect of a legal duty is not a crime unless some statute so prescribes; hence, if one were to drive a vehicle at night with the headlights turned off, unintentionally, and it were shown that the injuries would not have been inflicted had the lights been on as required by law, the court would not have to submit the question of negligence to the jury, but could rule that such omission was negligence per se, and in common law states, it would be criminal negligence.

nolle prosequi. An agreement not to proceed further in the prosecution as to a particular person or cause of action; the power to enter such a plea usually rests with the prosecuting attorney, but in some jurisdictions the consent of the court must be obtained.

nolo contendere. A plea which a defendant in a criminal prosecution may have entered for him, and amounts to an admission of guilt without his actually pleading "guilty."

nonfeasance. The failure to perform a required duty. See p. 8.

notices. See p. 123.

nuisance. Anything that works an injury to the individual or the public.

nuisance per accidens. Conditions which by themselves do not ordinarily constitute a nuisance, such as a house, but which if broken down, unsanitary, a fire menace, etc., might become one under certain circumstances. See p. 117.

nuisance per se. Any condition which at all times and under all circumstances, irrespective of location or surroundings, amounts to a nuisance. See p. 117.

ordinances. See p. 130.

organic law. That which is basic; the constitution of the government. See p. 1.
original jurisdiction. Jurisdiction of a court which is not appellate. See p. 16.
overtime. See p. 177.
patent defect. One which is clearly visible or obvious; not latent.
pension. See p. 160.
per se. As such; through itself. See p. 129.
plaintiff. A person who brings a suit. See p. 6.
pleadings. The allegations made by the parties to a civil or criminal case, whether to support or to defeat it.
pleas in abatement. See p. 9.
pleas in bar. See p. 9.
pleas of confession and avoidance. See p. 9.
police power. See p. 113.
post. Latin for "after."
presumption. That which may be assumed without proof or taken for granted. In some states the presumptions are set forth in statutes, and are designated as either conclusive or rebuttable; for example the law conclusively presumes that a child born in wedlock is legitimate where there has been opportunity for access and the husband is not impotent. It is also conclusively presumed that a person does an act with malice if he does it deliberately for the purpose of injuring another.

On the other hand, the presumption may be disproved that a person is innocent of crime or wrong, or that the money paid by one to another was due the latter, or that former rent or installments have been paid when a receipt for the latter is produced, or that official duty has been regularly performed, or identity of person from identity of name, or that a person not heard from in seven years is dead, or that a letter duly directed and mailed was received in the regular course of the mail, or that a writing is truly dated, or that the law has been obeyed. But unless the above presumptions and any other various or similar disputable presumptions are controverted by other evidence, the jury is bound to find according to the presumption. See p. 127.
prima facie. That which is made first; a first showing; e.g., that is prima facie correct which is presumed to be correct until the presumption is overcome by evidence which clearly rebuts it. Thus, prima facie speed limits set up by statute establish the maximum speed which will be deemed to be safe for ordinary conditions, and does not preclude arrest for going at said speed where the conditions are so hazardous that a slower speed would be required for safe operation. See p. 126.
prima facie case. See pp. 126, 205.
principal. In civil law, it is one who employs an agent to transact some business for him with a third party; it also refers to a person who is primarily liable for the performance of an obligation for which a surety is secondarily liable under the law of suretyship. In criminal law it is anyone who is present at the commission of a crime and who participates in it either directly or indirectly. See p. 13.

probable cause. Reasonable cause, or reasonable grounds. There is said to be probable cause for arresting a person on a felony charge without a warrant (so as to ward off a suit for malicious prosecution or false arrest) when there is a suspicion founded upon circumstances sufficiently strong to warrant a man in the belief that the charge is true. See pp. 8, 149.
probation. See pp. 47, 196.
procedure. See p. 20.
promissory estoppel. See "estoppel."
promotion. See p. 53.
proprietary function. A nongovernmental function, or corporate function (q.v.) when used with reference to a municipal corporation. See pp. 63, 76.
proximate cause. That cause of an injury which, in natural and continuous sequence, unbroken by any efficient intervening cause, produces the injury, and without which the injury would not have occurred. See p. 165.
public officer. See p. 35.
quasi. Relating to or having the character of; as if.
quasi judicial. The acts of an officer which are executive or administrative in character and which call for the exercise of that officer's judgment and discretion are not ministerial acts, and his authority to perform such acts is quasi judicial. See p. 192.
qui tam. An action brought by a nonofficial informer to recover a statutory penalty, one part of which goes to the sovereign or some public cause, and the other to the informer. See p. 134.
quo warranto. Latin for "by what authority." The proceeding whereby the right of a person to hold an office can be contested.
rebuttable presumption. See "presumption"; one which can be disputed. See p. 127.
removal. See p. 188.
res adjudicata. A matter which has been definitely settled by a judicial decision. See p. 174.
res ipsa loquitur. "The thing speaks for itself." See pp. 125, 126, 127, 209.
respondeat superior. "Let the superior respond"; let the master be responsible for the acts of his servant, or the principal for the acts of his agent. The doctrine requires that the individual higher in authority be responsible for the lower. See pp. 63, 219.
respondent. The party which is adverse to the appellant in an action which has been appealed to a higher court. See p. 6.
retroactivity. See p. 118.
rule of absolute liability. See discussion on p. 125.
rules and regulations. See p. 56.
search and seizure. See pp. 149, 232.
sequestration of witnesses. The ordering of them by the court to be separated or kept outside the court room until time for their testimony. See p. 240.
slander. See p. 6.
stare decisis. "Let the decision stand"; the doctrine that the decisions of the court should stand for future guidance, and all subsequent decisions on

the same issue should be bound by the precedent. See p. 202.

statute of limitations. The statute which sets up the respective periods of time within which various actions or proceedings must be instituted. For ex-example, in California, an action must be commenced upon a bond or note issued to the public within six years; four years, for any action founded upon certain contracts or instruments in writing, such as a trust deed or mortgage; three years for trespass or injury to real property; one year for a civil action founded upon libel, slander, assault and battery, false imprisonment, or injury caused by the negligence of another; ninety days to recover personal property left at a hotel, hospital, rest home, etc.; sixty days for a civil action based upon criminal conversation (where husband sues the man who had intercourse with his wife), or alienation of affections, seduction of a person over age of consent, etc. California has the following limitations on the filing of a criminal complaint against the accused: no limitation for murder, embezzlement of public moneys or falsification of public records; for a felony other than the above, three years with the exception of bribery, which is six years; misdemeanors, one year. See p. 165.

statutory law. See p. 2.

statutory liability. See pp. 71, 75, 109.

subrogation. The substitution of one person in the place of another with reference to a lawful claim or right, and frequently referred to as the doctrine of substitution; e.g., after paying the insured landlord the loss from a fire negligently caused by a tenant, the insured company has the right of subrogation to the claim which the landlord has against the tenant, and can sue the tenant to recover the amount paid out.

suit. Any proceeding in a court.

tort. An injury or wrong committed either with or without force, to the person or property of another; a civil injury not arising out of contract. See p. 8.

trespasser. A person who goes upon the premises of another without invitation, either express or implied, nor who is suffered or tolerated there. It is not necessary that the trespasser, in making the unlawful entry, should have an unlawful intent. Trespass is a tort, and unless it is accompanied by some other wrongful act, e.g., entering a posted area and building a fire, or hunting thereon, or tearing down a fence, or continuing to occupy the land after entering upon it, some states do not make the simple act of walking across another's property a crime. See p. 124.

ultra vires. "Beyond the scope of power or authority"; used chiefly with reference to a municipal corporation. See pp. 63, 165, 215.

unwritten law. The common law is often referred to as the unwritten law as distinguished from the statute or written law. See p. 2.

venue. Refers to the county or judicial district in which a case can be tried; determined by statute in most jurisdictions. See p. 8.

verdict. The answer of a jury given to a court concerning the matters of fact given to them to try and examine. See pp. 21, 241.

vested right. See p. 173.

warrants. See pp. 149, 233.
wilful misconduct. See p. 81.
workmen's compensation. See p. 182.
writ. A mandatory precept issued by the authority and in the name of the sovereign or the state, though issued by a court, for the purpose of compelling the person to do something or not to do something; e.g., writ of habeas corpus (q.v.). See p. 17.

APPENDIX A

Visitor's Waiver

TO: Chief Engineer, _____ Fire Department
FROM: _____

Request is hereby made for permission to ride on City of _____ Fire Department apparatus.

In consideration of your granting such permission, I agree to and do hereby waive any and all claims against the City of _____ or the Fire Department, or the members of the Fire Department with whom I ride, for any injury or accident occurring to me while on the Fire Department apparatus. This waiver shall be and is binding upon me, my heirs and personal representatives.

SIGNED: _____

DATE: _____

For Visitor File

APPENDIX B

Visitor's Identification

_____
Date

TO WHOM IT MAY CONCERN:

    The Chief of the _____ Fire Department has granted permission for me to work with various units and members of his Department. In consideration hereof, I have waived all claims of liability against the City of _____, the Fire Department or any member thereof for personal injuries I might incur while in or adjacent to Fire Department quarters and facilities or while riding or being on Fire Department apparatus. It is further understood that my presence in these various locations is granted subject to the discretion of officers of the Fire Department, and may be summarily revoked at any time.

    This permission is valid from _____ to _____.

_____
Signature

_____
Type Applicant's Name

_____
Approved by

_____
Rank   Assignment
Orig: Administrative Office
Dupl: To Applicant

APPENDIX C

California Law on Liability of Public Entities and Employees

Background

In drafting the new law relating to governmental liability the California Legislature recognized the need for continued tort immunity in the field of fire suppression and prevention. Although few cases have been decided under the new Government Code Sections on this subject thus far, the traditional immunity cited in recent California decisions holding a city not liable for damages due to fire where the failure of the fire department to extinguish the blaze was the consequence of negligent failure to properly maintain the fire hydrants and water mains in workable condition (38 Cal. 2d 486), or to provide reasonably adequate fire fighting equipment (21 Cal. Rptr. 398), may undergo modification in future cases. However, the Modesto case, discussed earlier in the text but decided since the new code provisions were adopted, did not upset the traditional concept of municipal immunity.

Most of the decisions which related to torts arising out of fire fighting activities did not attempt to analyze the basis for the immunity doctrine, but unquestioningly applied the traditional rule. Once in a while a court would deny recovery on the grounds that to do otherwise would have disastrous fiscal consequences arising from liability. In New York, for example, where the law (New York Ct. Cl Act. Section 8) imposes liability on cities the same as on individuals, the courts nevertheless held there was no liability for the negligent failure to maintain adequate water pressure for fire fighting purposes because an opposite result might impose a "crushing burden" on the city (295 N.Y. 51).

In Florida, even though the courts have disapproved of the governmental immunity doctrine (96 So.2d 130), a recent decision refuses to impose tort liability on a city for negligence in fire fighting activities of its fire department on the ground that "a conflagration might cause losses, the payment of which would bankrupt the community" (132 So.2d 764). Similar views have been widely expressed elsewhere, such as in the case of the City of Columbus v. McIlwain (205 Miss. 473), where the court earlier declared, "If such liability existed, history records many disastrous fires which would have resulted in the complete bankruptcy of the municipality." Cases along the same line include those for Illinois, Indiana, Ohio and New Hampshire. The Supreme Court of Tennessee also warned that "the hazard of pecuniary loss," if the immunity doctrine was abrogated as to fire fighting, might become so great as to "frighten our municipalities from assuming the startling risk" (101 Tenn. 291).

While it's true that there might be some fiscal consequences arising from a policy of unlimited liability of public bodies engaged in fire fighting, yet this should not be the controlling factor in developing a rational rule of law. This general rule of immunity, which is predominant throughout the east coast and middle west of the United States, and which has characterized the law of California before the Muskopf decision, seems to have expressed fiscal concern primarily for the government entity. Yet we all know that serious conflagrations happen today as they have in the past, as for example, the Bel Air disaster which occurred in the author's fire suppression command in November of 1961 in which almost 500 homes in the brush-covered foothills of Los Angeles were destroyed in a single day.

Therefore, in addition to the effect on the community, we also need to consider the disastrous consequences to the individuals who lose their homes and all their possessions due to the negligently caused fires which are beyond the power of the fire department to prevent or suppress. And even though we have this concern over the loss of the individual, this does not mean that the government must become an insurer against any and all fire losses for its liability should only be imposed where there is *fault*, and in most large conflagrations the losses to the individual property owner could seldom be proved to arise out of any negligence in fire fighting or other misconduct on the part of the fire fighting forces.

On the other hand, it is conceivable that the fire department might be negligent in handling a fire in its incipiency—where, while it is still small, it could have been controlled to prevent its spread and to prevent it from developing into a serious disaster; in such a case the imposition of liability might be justified, but some limitation on that liability would have to be worked out to keep it within reason. For example, if the liability were limited to only the damage that was "proximately caused" by the negligence, then under circumstances where intervening conditions might arise to extend the losses beyond that which might reasonably be expected to follow such negligence, the liability could be cut off. If hot weather, high winds and low humidity were present, and especially in a brush-covered, mountainous terrain, having dead-end streets and a water supply dependent upon elevated storage tanks whose replenishment would be cut off by fallen power lines which supplied the electric motors for the tanks' pumps, then certainly these factors could be considered to have intervened to cut off the chain of normally anticipated consequences of the fire fighter's possible negligence in failing to promptly confine the fire in the first place. Under such circumstances the entity might perhaps be held liable for negligence which results in the destruction of the property which initially catches on fire and possibly for the loss of the immediately adjoining property which likewise would forcibly be exposed to the loss by any such negligence; but provided reasonable action is taken in the light of the whole problem to prevent the further spread of the blaze, no liability should attach for private losses beyond the initial perimeter of the fire.

Another reason that was frequently given for the adoption of the rule against liability in fire fighting was the one that since the city was not required to provide fire protection in the first place, it should not be held liable if it does so poorly or negligently.

Since fire departments are now considered to be rendering one of the most basic and fundamental services which the government offers its citizens, its maintenance can no longer be considered "voluntary," except in the broad sense that all organized government is a voluntary undertaking, and the question of tort liability of its officers and employees should be considered on the basis that they are performing a public duty and not merely a philanthropic benefit.

The removal of governmental immunity is actually not new, for as the court pointed out in the Muskopf case, complete immunity has never existed. As a matter of fact, the underlying law is that liability *does* arise where negligence is involved; immunity, whether governmental or any other kind, is the historical exception, and in the United States, insofar as fire service activities are concerned, it has generally related to the non-liability of public agencies for the operation of fire trucks and other fire fighting equipment, or to dangerous and defective conditions of public property under the jurisdiction and control of fire departments.

In California, however, liability has long been imposed in these two fields by section 17001 of the Vehicle Code, relating to the operation of fire trucks, and section 53051 of the Government Code, relating to defective fire department property.

Other California statutes make selected public entities liable for the torts of their personnel while engaged in carrying out fire suppression and prevention functions, chiefly in the form of provisions requiring the entity to satisfy any judgments against its personnel arising out of torts committed by them in the course of their duties, e.g., Government Code, Sec. 61633, requiring community services districts to satisfy judgments against their personnel arising out of torts committed in the course of fire suppression and prevention duties of such districts. Government Code, Sec. 39586, 53057, imposes liability for personal injury and property damage from burning of weeds and rubbish.

The Law—Specifically

Now that we have reviewed the background for the law in California relating to the liability of a public entity, let us examine more closely some of the specific sections which will be of interest to the fire service. The sections with which we are primarily concerned commence with Government Code section 815.2 which makes the city liable for any injury caused by any of its employees who are acting within scope of their employment. This means that the old doctrine of "Respondeat Superior" has been revived. It does away with any distinction between "governmental" and "proprietary" functions. It includes the state and other governmental agencies in the liability, which is much broader than the old Public Liability Act of 1923 where only the city, county and school districts were included.

In section 815.6, it provides that where a public entity is under a mandatory duty imposed by law that is designed to protect against the risk of a particular kind of injury, then it becomes liable for an injury of that kind caused by its failure to discharge the duty, unless it can establish that it exercised reasonable diligence to discharge the duty. In other words, if a fire breaks out and the fire department is notified, a city could be held liable if its fire fighters fail to exercise reasonable diligence to extinguish the fire, since it has a mandatory duty to do so.

The law also provides that a public entity is not liable for an injury caused by adopting or failing to adopt an enactment or by failing to enforce the law. Also, that it is not liable for an injury caused by the issuance or denial or suspension or revocation or the failure to refuse, issue, deny, suspend or revoke any permit, license, certificate, etc., where the employee is authorized by law to determine whether or not such permit should be issued, denied, suspended, or revoked. These exemptions would not have been necessary if section 815.6, previously discussed, had not been adopted, making the government liable where there is negligence in failing to carry out a mandatory duty. In other words, where there is a discretionary power, now there will be no government liability, which is an enactment into law of what was decided in the Lipman case.

Another section of the Government Code provides that the city will not be liable for an injury caused by its failure to make an inspection or by reason of making an inadequate or negligent inspection of any property (other than its own property, which is covered by Section 830), such inspections being made for the purpose of determining whether or not the property complies with or violates any enactment or constitutes a hazard to health or safety. Here again it is a discretionary power for which the city will not be liable if it is not properly exercised; if this immunity were not granted, the city would be held liable for virtually all property defects within its jurisdiction. But keep in mind that it does not excuse the city from inspecting its own property, nor would it prevent an inspector from becoming liable for active negligence, or for assault and battery, or other wilful misconduct on his part while performing his inspections.

In section 818.8 the public entity is not liable for any injuries caused by any misrepresentation by any employee, whether it is negligent or intentional.

Except as elsewhere provided by statute (e.g. Vehicle Code section 17004), a public employee is liable for injury caused by his act or omission to the same extent as a private person, and the liability is subject to any defenses that would be available to a public employee if he were a private person. However, except as otherwise provided by statute, he is not liable for an injury resulting from his act or omission where the act or omission was the result of a *discretion* vested in him, whether or not such discretion be abused. He is not liable for any act or omission exercising due care in the execution or enforcement of any law, but of course nothing in this law will exonerate him from liability for false arrest or false imprisonment.

Section 820.6 states that if a public employee acts in good faith without malice, and under the apparent authority of an enactment that is unconstitutional, invalid or inapplicable, he is not liable for an injury caused thereby except to the extent that he would have been liable had the enactment been constitutional, valid and applicable. This law actually broadens the former law which only allowed the man immunity in the event that the law turned out to be *unconstitutional;* this extends it to *invalidity* and *inapplicability*

Section 820.8 provides that a public employee is not liable for an injury caused by the act or omission of another person, unless otherwise provided by statute, and of course nothing in this section exonerates the public employee from liability for injuries proximately caused by his own negligence or wrongful acts or omissions. This nullifies many of those old cases which held that in some instances an employee could be held liable for the acts of his subordinates. Of course he can still be held liable for the acts of his subordinates if he encourages them to do something which injures another, or if he knows that they are engaging in wrongful acts and does nothing about it, or if he knows they have committed a tort and approves of it or ratifies it; in such cases he would still be liable. But as a general rule, you now can state that a man is only liable for his own torts and not those of another person, even if he works for him.

Section 821 declares that a public employee is not liable for an injury caused by his adoption of or failure to adopt an enactment or ordinance, or by his failure to enforce such an enactment. For example, there would be no liability for failure to arrest a drunk driver as indicated in the case of Rubinow v. the County of San Bernardino (169 Cal. App.2d 67, 1959).

In Section 821.2 it states that a public employee is not liable for an injury caused by his issuance, denial, suspension or revocation of, or by his failure or refusal to issue, deny, suspend or revoke, any permit, license, certificate, approval, order, or similar authorization where he is authorized by enactment to determine whether or not such authorization should be issued, denied, suspended or revoked. This has been a long established immunity in California (Downer v. Lent, 6 Cal. 94, 1856; pilot commissioners' immunity from liability for maliciously revoking pilot's license).

In Section 821.4, it states that a public employee is not liable for injury caused by his failure to make an inspection, or by reason of making an inadequate or negligent inspection, of any property, other than the property of the public entity employing the public employee, for the purpose of determining whether the property complies with or violates any enactment or contains or constitutes a hazard to health or safety. This relates only to the adequacy of the inspection and does not provide immunity if he injures a person while he is making the inspection. A fire fighter on night hotel inspections would be immune from *civil* liability if he failed to observe a padlocked exit, but not be immune from possible *criminal* liability, for example where the locks on the doors subsequently prevented guests from escaping from a fire in that hotel. Also, note that this immunity would not exempt him from liability if the dangerous condition which he failed to observe was in his own fire station, and the injured victim was successful in his lawsuit. (See section 840 to 840.6.)

Section 821.6 states that a public employee is not liable for injury caused by his instituting or prosecuting any judicial or administrative proceeding within the scope of his employment, even if he acts maliciously and without probable cause.

Section 821.8 provides that a public employee is not liable for an injury arising out of his entry upon any property where such entry is expressly or impliedly authorized by law. Nothing in this section exonerates a public employee from liability for an injury caused by his own negligent or wrongful act or omission. This nullifies the old common law rule that a public employee who enters a property under authority of law, but then commits a wrongful or negligent act, is a trespasser ab initio and liable for all damages resulting from his entry.

A public employee is not liable for money stolen from his official custody. Nothing in this section exonerates a public employee from liability if the loss was sustained as a result of his own negligent or wrongful act or omission (Section 822). Nor is a public employee acting in the scope of his employment liable for an injury caused by his misrepresentation, whether or not such misrepresentation be negligent or intentional, unless he is guilty of actual fraud, corruption or actual malice. (See 822.2.)

Section 825 of the new Government Code provisions deals with the idemnification of public employees. If an employee or former employee of a public entity requests the public entity to defend him against any claim or action against him for an injury arising out of an act or omission occurring within the scope of his employment as an employee of the public entity and such request is made in writing not less than 10 days before the day of the trial, the public entity shall pay any judgment based thereon or any compromise or settlement of the claim or action to which the public entity has agreed. The sections in this Article, Article 4, require public entities to pay claims in judgments against public employees that arise out of public employment. A city, for example, could only recover from the employee where the latter has acted with actual malice, actual fraud or corruption. But to avoid conflict of interest, the city waives its right to recover from the employee if it furnishes his defense without a reservation of its rights not to pay the judgment.

In Chapter 2 of this division of the Government Code, the court decisions relating to dangerous conditions of public property are embodied in the law. It does, however, try to avoid giving rise to liability for trivial defects such as happened in the sidewalk cases, where a lady might sue the city to collect damages for a fall that occurred when her high heel slipped into a crack that wasn't over a quarter-inch deep and 10 inches long.

Section 830.2 states that: A condition is not a dangerous condition within the meaning of this chapter if the trial or appellate court, viewing the evidence most favorable to the plaintiff, determines as a matter of law that the risk created by the condition was of such a minor, trivial or insignificant nature in view of the surrounding circumstances that no reasonable person would conclude that the condition created a substantial risk of injury when such property or adjacent property was used with due care in a manner in which it was reasonably foreseeable that it would be used.

Another section provides that there is no liability for failure to install traffic signals, stop signs, yield right-of-way signs, etc. It also provides that just the mere fact that there was an accident doesn't necessarily mean that the public property was in a dangerous condition, or the fact that remedial action was taken after the injury occurred is not necessarily evidence that the public property was in a dangerous condition at the time of the injury.

Section 830.6 exempts the city from liability for an injury caused by a plan or design of construction or improvement of public property where the design had been approved in advance by the legislative body of the entity or some other body exercising discretionary authority to give approval where the plan is prepared in conformity with standards previously so approved, providing the court finds there is substantial evidence upon which it can be found that a reasonable public employee could have adopted the plan or design and a reasonable legislative body could have approved the plan or design or the standards therefor.

Section 830.8 provides that the city can be held liable if it fails to put up a warning sign where a dangerous condition exists such as a ditch or excavation which would not be reasonably apparent to a person exercising due care.

In Article 2, on the subject of "Liability of Public Entities," the Government Code provides that, except as provided by statute, a public entity is liable for injury caused by a dangerous condition of its property (note that this includes any kind of property—real or personal) if the plaintiff establishes that the property was in a dangerous condition at the time of the injury, that the injury was proximately caused by the dangerous condition and that the dangerous condition created a reasonably foreseeable risk of the kind of injury which was incurred, and that either: (a) a negligent or wrongful act or omission of an employee of a public entity within the scope of his employment created the dangerous condition, or (b) the public entity had actual or constructive knowledge of the dangerous condition a sufficient time prior to the injury to have taken measures to protect against the dangerous condition.

835.2 provides that (a) a public entity had *actual* notice of a dangerous condition if it had actual knowledge of the existence of the condition and knew or should have known of its dangerous character, and (b) a public entity had *constructive* notice of a dangerous condition within the meaning of the above section only if the plaintiff established that the condition had existed for such a period of time and was of such an obvious nature that the public entity, in the exercise of due care, should have discovered the condition and its dangerous character. The law also covers the meaning "due care"; for example, whether the condition was one which would have been discovered by an inspection system that was reasonably adequate considering flexibility, costs, and so forth, and whether or not the public entity maintained and operated such an inspection system with due care and did not discover the condition.

835.4 provides that the city is not liable for an injury caused by a condition of its property if the public entity establishes that the act or omission that created the condition was reasonable. For example, would it be reasonable to build a reservoir over a known geological fault where the subsequent slippage of the land might result in the dam breaking and many homes being washed away? The reasonableness of the act or omission that created the condition would be determined by weighing the probability and gravity of potential injury to persons and property foreseeably exposed to the risk of injury against the practicability and cost of taking alternative action that would not create the risk of injury, or of protecting against the risk of injury. As applied to the dam case, a jury might have to weigh the question of whether or not it would cost so much and would be so impractical to put a needed water storage capacity in hundreds of elevated tanks, for instance, in lieu of using a reservoir, and/or whether or not the risk of earth slippage was so remote that it was reasonable to expose the public to this threat of damage to buildings in the event the dam should rupture by constructing the reservoir on top of a hill over geological formations which were known to slip occasionally.

This section also provides that the city would not be liable for injuries caused by a dangerous condition of its property if it establishes that the action it took to protect against the risk of injury created by the condition or its failure to take such action was reasonable. The reasonableness of the action or inaction is determined by taking into consideration the time and opportunity it had to take action and by weighing the probability and gravity of potential injury to persons and property foreseeably exposed to the risk of injury against the practicability and cost of protecting against the risk of such injury.

Article 3 of the Government Code relates to "Liability of Public Employees," starting with Section 840. It states that ". . . Except as provided in this article, a public employee is not liable for injury caused by a condition of public property where such condition exists because of any act or omission of such employee within the scope of his employment. The liability established by this article is subject to any immunity of the public employee provided by statute and is subject to any defenses that would be available to the public employee if he were a private person."

Section 840.2 provides that an employee of a public entity is liable for injury caused by a dangerous condition of public property provided that it was in a dangerous condition at the time of the injury, that the injury was proximately caused by the dangerous condition, that the dangerous condition created a reasonably foreseeable risk of the kind of injury which was incurred and that either the condition was directly attributable wholly or in part to the negligent or wrongful act of the employee and the employee had the authority and the funds and other means immediately available to take alternate action which would not have created the dangerous condition; or the employee had the authority and it was his responsibility to take adequate measures to protect against the dangerous condition at the expense of the public entity and the funds and other means of doing so were immediately available to him, and he had actual or constructive notice of the dangerous condition a sufficient time prior to the injury to have taken measures to protect against it.

840.6 states that the employee is not liable for an injury caused by a dangerous condition of public property if he establishes that the act or omission that created the condition was reasonable. The reasonableness of the act or omission that created the condition shall be determined by weighing the probability and gravity of potential injury to persons and property foreseeably exposed to the risk of injury against the practicability and cost of taking alternative action that would not create the risk of injury or of protecting against the risk of injury. This would apply, for example, where it was reasonably possible during Fire Service Day to protect pole holes against the possibility that children might fall through them while wandering around while on the upper floors; certainly it would not be considered impractical to wrap fire hose around the guard rails of these openings to prevent small children from inadvertently falling through them; not to protect them at all would undoubtedly be considered negligent.

The reasonableness of the inaction or action is determined by taking into consideration the time and opportunity the public employee had to take action and by weighing the probability and gravity of potential injury to persons and property foreseeably exposed to the risk of injury against the practicability and cost of protecting against the risk of such injury. This section might apply in a case where a fire fighter fails to set the brakes on his truck or to cut the wheels into the curb, and the truck subsequently rolls down the hill and smashes into somebody's home or car. The question might arise whether or not the failure to place the chock blocks or to set the brakes or to do both was negligence in view of the time he had available or the opportunity to take these precautions and whether or not he should have reasonably foreseen that the results of his failure would cause serious damage. It is easily conceivable that a jury might find that failure to take such precautions under conditions where ample time was available would be considered lack of due care.

A special chapter relates to fire protection in the Government Code—Chapter 4; it commences with Section 850, which states that neither a public entity nor a public employee is liable for failure to establish a fire department or otherwise to provide fire protection service.

850.2 states that neither a public entity that has undertaken to provide fire protection service, nor an employee of such a public entity, is liable for any injury resulting from the failure to provide or maintain sufficient personnel, equipment or other fire protection facilities.

The above sections (850 and 850.2) were held not a bar to liability of a county which had a dangerous condition (gasoline and chemicals) on its property, and had provided no fire protection for it. The court said that immunity statutes must be strictly construed (Vedder v. Imperial County, 36 Cal. App.3d, 1974).

850.4 provides that neither a public entity, nor a public employee acting in the scope of his employment is liable for any injury resulting from the condition of fire protection or fire fighting equipment or facilities or, except as provided in the Vehicle Code, for an injury caused in fighting fires. This section grants fire fighters the broad immunity which they need, and which the courts have long recognized they have to be given in order to effectively accomplish their tasks. This immunity, however, does not exist if there is a wilful or negligent failure to fight a fire, for fire fighters will still be expected to use the same care that an ordinary prudent fire fighter would use under like circumstances. Nor does this immunity extend to training activities, nor does it excuse them from liability which might arise from the negligent operation of motor vehicles. It does relieve the fire fighters and the city of liability for injuries resulting from lack of equipment or faulty equipment or the fire fighting equipment which isn't suitable for the needs.

Section 850.6 relates to providing assistance to another community. Whenever a city, upon receiving a call for assistance from another public entity, provides fire protection or fire fighting service outside of the area which it regularly serves, the city providing such service is liable for any injury for which liabilty is imposed by statute caused by its act or omission or those of its employees occurring in the performance of such fire protection or fire fighting service. But, the public entity calling for assistance is not liable for any act or omission of either the city providing the assistance or of its employees. However, the city providing such service and the public entity calling for assistance may, by agreement, determine the extent, if any, indemnification will be required.

Lastly, Section 850.6 provides immunity to firemen who transport injured persons from a fire to a hospital, and renders them liable only for wilful misconduct.